CAMBRIDGE STUDIES
IN MATHEMATICAL BIOLOGY: 3

Editors
C. CANNINGS
Department of Probability and Statistics, University of Sheffield

F. HOPPENSTEADT
Department of Mathematics, University of Utah

GENEALOGICAL AND GENETIC STRUCTURE

T0245350

C. CANNINGS

Department of Probability and Statistics, University of Sheffield

E. A. THOMPSON

Department of Pure Mathematics and Mathematical Statistics
University of Cambridge

Genealogical and genetic structure

CAMBRIDGE UNIVERSITY PRESS

Cambridge

London   New York   New Rochelle

Melbourne   Sydney

CAMBRIDGE UNIVERSITY PRESS
Cambridge, New York, Melbourne, Madrid, Cape Town, Singapore, São Paulo, Delhi

Cambridge University Press
The Edinburgh Building, Cambridge CB2 8RU, UK

Published in the United States of America by Cambridge University Press, New York

www.cambridge.org
Information on this title: www.cambridge.org/9780521239462

© Cambridge University Press 1981

First published 1981
Re-issued in this digitally printed version 2008

A catalogue record for this publication is available from the British Library

Library of Congress Catalogue Card Number: 81–6100

ISBN 978-0-521-23946-2 hardback
ISBN 978-0-521-28363-2 paperback

CONTENTS

PREFACE

This monograph is an attempt to provide both an introduction and an overview of the importance of, and the techniques for handling, *relationships* in a genetic context. It is hoped that both graduate students and research workers will benefit from the material presented, whether specialists or workers from other fields.

Much of the material can be appreciated without sophisticated mathematics, our object being to present methods and results rather than detailed proofs. We have, where possible, indicated open questions and possibly fruitful lines of advance, and attempted to give copious though not exhaustive background references.

The book describes the various ways in which we may represent, quantify and use relationships. The first chapter specifies one way in which relationships may be represented, and how one may reduce compound relationships to simpler ones. The second chapter introduces the concept of genetic identity, and this is exploited in the context of genealogies, and, in Chapter 3, of populations. Chapter 4 considers the genetic variability within a population, and Chapter 5 discusses measures of genetic distance between individuals, genotypes and populations. The final chapter is devoted to setting out methods for calculating the measures developed earlier in the book.

We would like to express our appreciation to Joe Felsenstein for comments on the manuscript, and to Hazel Howard, Rosemary Moreton, Rita Perry and Fiona Carter for their careful typing of it.

Finally we acknowledge an indebtedness to Mark Skolnick whose support during visits by the two authors to his group in the Department of Biophysics and Medical Computing, at the University of Utah, Salt Lake City, has helped and encouraged us to complete the present work.

C.C.
E.A.T.
September 1980

GLOSSARY

Symbols are here listed in alphabetical order, with script letters preceding upper case letter, which precede lower case letters. Greek symbols are listed at the point appropriate for their English name. Symbols with sub- and superscripts are treated as a *word* for purposes of ordering, subscripts preceding superscripts. All subscripts are given as i or i, j; thus A_j would be looked up under A_i. Non-alphabetic symbols are listed at the end. Some symbols used only within a short section of text, and there defined, are not included.

Symbol	Meaning	Chapter or main section(s)
$\mathscr{A}$	allocation of allelic types to genes	2.5
A	a specified allele	*passim*
A_i	ith allele	*passim*
A_{ij}	$P(i$ genes contain j alleles$)$	4.3
a_i	additive effect of allele A_i	2.6
$a_i(t)$	$P(i$ alleles at time $t)$	4.3
$a(t)$	number of alleles at time t	4.3
α	inbreeding coefficient	*passim*
$\alpha(B)$	inbreeding coefficient of individual B	*passim*
B	as subscript; between groups	3.9, 4.4
B	a specified individual	2
$\{B_i\}$	a set of specified individuals	2.2
B^*	a specified individual	2.2
β	mutation/migration rate	3.4, 4.5
$C_{i,j}$	jth child in ith sibship	5.3
D_i	number of identities between i individuals	2.5, 2.7
d	measure of distance	5.4
d_{ij}	dominance effect of genotype A_iA_j	2.6
$d_i, d_k, d_{\mathrm{LM}}$	measure of distance	5.4

E	expectation	*passim*
$E(B, B^*, \ldots)$	event that phenotypes of B, B^*, ... are as observed	6.5
F	father (when denoting an individual)	2.2, 5.3
F	realized process of gene identity by descent	4.5
F_i	father of ith sibship	5.3
F_{ST}	allelic correlation within a subpopulation	4.4, 4.5
f	as subscript; female/maternal	2
$\mathbf{G}$	genotypic structure	4
Γ_{ij}	$P(i$ genes from j parent genes)	3.5, 4.3
$G_B(i)$	event that individual B has the ith genotype	6.5
$\mathbf{G}(\mathrm{HW})$	Hardy–Weinberg genotypic structure	4.2
$\mathbf{G}(I)$	fully-inbred genotypic structure	4.2
G_i	genotype of ith individual	2.4
G_{ST}	generalization of F_{ST}	4.5
γ	mutation/migration rate/probability of gene descent	3.4, 2.7
g_i	ith gene	2.4
$g(t)$	number of genes at time t	4.3
H	non-identity of descent, or type/process of non-identity of type	3, 4.5
HS	half-sibs, each produced by sexual mating	2.2, 3.7
HS*	half-sibs, one produced by sexual mating and the other by selfing	2.2, 3.7
HW	Hardy–Weinberg	*passim*
I	identity of homologous genes	*passim*
IBD	identical by descent	2
i	general subscript	
J	identity of allelic type	4.5
j	general subscript	
K	array of identity specification of population	2.3, 4.5
K	total numbers of elementary neighbourhoods	4.4
$\mathbf{k} = (k_2, k_1, k_0)$ probabilities of genes IBD		2.3
k	general subscript	
L	likelihood	5.2, 5.3
λ	latent root of gene identity recursions	3.2
M	marriage	1
M	mother (when denoting an individual)	2.2, 5.3
M_i	mother of ith sibship	5.3
m	as subscript; male/paternal	2

m	general integer (e.g. mth cousins)	
μ	mean	2.6, 3.7
N	population size (individuals or genes)	*passim*
N_e	effective population size	3
n	neighbourhood or group size	2.4, 4.4
n_i	size of ith subpopulation	4.4
$\mathscr{P}$	population	5.4
P	probability	*passim*
PO	parent-offspring, offspring produced by sexual mating	2.2, 3.7
PO*	parent-offspring, offspring produced by selfing	2.2, 3.7
P_{ij}	probability that i parental genes give rise to j offspring genes	4.3
π_i	probability of ith gene identity state	2, 5.4
p_i	population frequency of allele A_i	*passim*
ϕ_i	phenotype of ith individual (viz. B_i)	2.5, 5.2
ψ	kinship coefficient	*passim*
$\psi(B, B^*)$	kinship coefficient between B and B^*	*passim*
ψ_W	kinship within groups	4.4
$Q_i^{(t)}$	probability of i surviving genes at time t	4.3
$q_i^{(t)}$	probability of extinction of specified set of i genes by time t	4.3
$\mathscr{R}$	a set of genealogical relationships	5.3
R	reproduction	1
R	a specified genealogical relationship	2, 3.7
R_{B,B^*}	matrix of conditional or joint probabilities for the genotypes of B and B^*	6.5
r	measure of relationship	5.4
S	sibs	2.2, 3.7
S	gene identity state	2.4
S*	sibs produced by selfing	2.2, 3.7
S	integer characterizing subpopulation size	4.4
ST	as subscript; subpopulation relative to the total	4.4, 4.5
s	number of subpopulations	4.5
σ^2	variance	3
σ_1^2, σ_2^2	variances of family size	3.7
$\sigma_B^2, \sigma_{BW}^2, \sigma_{WW}^2$	variances between and within groups	4.4
t	time and/or generation number	*passim*

$\mathcal{U}$	a set of null relationships (unrelated individuals)	5.3
V_A	additive genetic variance	2.6
V_D	dominance variance	2.6
$v_{B^*}(B)$	ancestral contribution of B^* to B	5.4
W	as subscript; within groups	3.9, 4.4
Z	log-likelihood ratio	5.2
○	an individual	1
●	a marriage	1
↠	a reproduction arc	1
→	a marriage arc	1
+	male/paternal	1.3, 6.3
−	maternal	6.3
*	female	1.3

1

Genealogical relationships

1.1   Introduction: the notion of relationship

This book is aimed at the geneticist, for whom genealogical relationships have a primary importance. The significance of these relationships is that they define the paths along which the hereditary material has been transmitted. We begin by outlining some of the problems which geneticists wish to solve, and leave aside for the moment a formal definition of relationships.

Consider firstly the medical geneticist. He is usually interested in a specific characteristic (or a small number of characteristics) within some specific group of individuals. He is concerned with four types of information:

(i) the genealogy (i.e. the set of relationships) for the group,

(ii) the genotypes (genetic types) and phenotypes (expression of the character) of the individuals,

(iii) the mechanism of transmission (possibly involving both hereditary and environmental components), and

(iv) the underlying population frequencies of the various types.

Information will be available on some, or all, of these categories, and the job of the medical geneticist is to fill in gaps in the information. The simplest types of problem are those of genetic counselling (Fraser Roberts and Pembrey, 1978), where it is required to predict the type of some potential offspring, and that of paternity testing (Chakraborty, Shaw and Schull, 1974), where one is trying to fill a gap in the pedigree. More complex problems often arise but these need not concern us here.

For the population geneticist the problems are somewhat different. His objective will be to explain the pattern of changes, which have taken place in the genetic structure of a population, and to predict future changes (possibly as the result of some proposed medical program). He is unlikely to possess a precise genealogy, and will have to confine himself to making statements about the features of randomly selected individuals (or groups), rather than about specific individuals. As we shall see in Chapter 3 his predictions will depend on the relationship structure of the population, and this will be influenced by the marriage system (i.e. the way individuals are chosen as spouses), and the demographic parameters. Ideally we would wish to take both the marriage system and the demographic parameters as subject to evolution, but for our present presentation we shall take them as immutable.

It is essential at this point to broaden our viewpoint, and to discuss relationships as they are seen by the social anthropologist. For him relationships are also of paramount importance, because they often determine, or strongly influence, the type of behaviour which an individual exhibits towards another, and expects from another (see, for example, Goody (1971)).

In human societies the acquisition of property, sovereignty, or other rank, on the death of its possessor is, more often than not, decided on the basis of relationship. The passage of rank, where there is a single, indivisible entity to hand on, will proceed by some fairly simple system; the system of primogeniture being the one used in recent times by the British monarchy and peerage. With the passage of property, which is potentially divisible, the possible systems are much more varied, and the system adopted will ultimately have consequences for the overall distribution of property within the society. Clearly, in a society where the ability to acquire wives, and hence children, depends on rank and/or wealth, these aspects of succession and inheritance will have profound demographic and genetic consequences. Neel (1970) has discussed the family structure of the Yanomama, where the system produces gross differences in the reproductive output of different males.

These considerations apply to a lesser extent to animal groups. The only property held is likely to be a territory, which if held by the group may remain with the groups descendants, but is unlikely to be passed on by an individual to his own descendents. On the other hand it has been demonstrated that, in some primate groups, an individual acquires the status of his mother. In Japanese macaques this occurs apparently by virtue of the involvement of mothers in fights between their offspring (Gray Eaton, 1976).

The obligations and rights of individuals towards each other during life are often dependent on relationship. In modern, western, human society, obligations to one's spouse, parents and children are common, while in more (technologically) primitive ones, there are often many and varied responsibilities, and rights, extending beyond close relatives. Levi-Strauss (1968) has argued that the most important right of all is the right to give a woman as a bride. The importance of this right rests in being able to obtain wives in exchange for those given, either for oneself, or for one's close relatives. Levi-Strauss further argues that this system leads to there being rules to prevent this right being eroded, as it would be if individuals could take a spouse from within their own group. Thus a rule of exogamy, and a ban on incest and marriage between close relatives, arises as secondary effects. The importance of the exchange also lies in its effect in forming and cementing alliances. Naturally these phenomena, of exogamy and the prohibition of incest (whether or not one accepts the anthropologist's explanation of their origin and persistence), expressed as they are through various marriage systems (Fox, 1967), are of great importance to the population geneticist. The system adopted to a large extent determines the frequencies of the various types of relative in the population, and the shape of the genealogies. Additionally, the frequently found prohibition of close relatives reduces the chance of certain genetic deficiencies, to which the children of closely related spouses are particularly exposed. We should note that while exogamy seems to be fairly widespread many groups adopt contrary systems (Mair, 1971). A system of endogamy (marriage within the group) has the advantage of keeping property within the group, and forming closely knit groups. Nonetheless exogamy is the commoner practice.

Prohibitions against incest in human societies seem to be parallel by effective avoidance in many animal groups. In discussing this question, Wilson (1975) gives reference to work on lions, old world monkeys and apes, mice, rats and guinea pigs, which indicates the existence of mechanisms to prevent, or reduce, the frequency of incestuous matings. Mechanisms to avoid selfing (i.e. mating with oneself) exist in many plants and hermaphrodite animals. Additionally, in plants, methods of pollen and seed dispersal may favour outbreeding.

We shall accommodate certain aspects of the above systems within our models, though one should keep in mind that these are necessarily simplifications of the underlying realities.

1.2   Genealogies

The complete representation of the relationships within a set of individuals is possible in various ways, each essentially involving the listing of every individual and his two parents, if these are within the group. Such a list constitutes a genealogy, and the usual diagrammatic form for representing such a genealogy is shown in Figure 1.2(*a*).

An alternative representation (for the same genealogy) is shown in Figure 1.2(*b*). Here we have used what is known as a digraph (directed graph). There are two types of node: ○ representing an individual, and ● representing a marriage, and these entities are linked by two types of directed arc: → joining an individual to a marriage of which he is a member, and accordingly called a marriage arc, and ↠ joining a marriage to an individual produced by that marriage, and accordingly called a reproductive arc. We will call this form of representation the graph of the genealogy. We believe that such a graph is a more useful representation of the genealogy than the usual form. The introduction of a node for each marriage facilitates the representation of relationships, *vis-à-vis* one working entirely in terms of individuals.

Figure 1.2(*a*). The usual diagrammatic form of representation of a genealogy. Here, for example, 1 and 2 are spouses, and have two offspring 5 and 6, the latter being the spouse of 7.

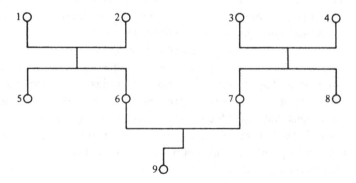

Within a genealogy each pair of individuals have a specific relationship to each other, which is defined by the chain (or chains) of arcs linking their nodes. For the moment we define some simple terms. Two individuals are said to be related if there is a chain joining them. The type of chain, and hence of relationship, can be divided into biological and affinal (through marriage). An affinal chain is one which contains a pair of consecutive marriage arcs, → followed by ←, a biological chain is one which contains no such pair. Two individuals are biologically related if there exists one, or more, biological chains joining them. One individual is said to be the ancestor of another if linked to the latter by a chain of alternating ← and ⇐ arcs. If A is an ancestor of B, then B is a descendant of A. Thus two individuals are biologically related if one is an ancestor of the other, or if they have a common ancestor. When two such individuals mate any offspring is said to be inbred (Fisher, 1949), and the appropriate representation within the graph will be similar to one of the cases in Figure 1.2(c): in each case a closed loop will occur. It should be noted that other types of loops, in which no inbreeding is involved, may occur (Cannings, Thompson and Skolnick, 1978).

Figure 1.2(b). The genealogical graph (see text for explanation). Note that each ● has two → arcs incident to it.

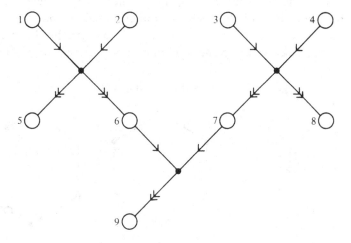

1.3   **Algebras**

We have seen how the complete set of relationships within a group can be represented. Often however we wish to discuss only the relationship between two individuals, or between one individual and several others, and it is useful to have available certain names and symbols for the more common of these.

Our aim here is to obtain a representation of certain basic relationships, without the necessity of using the graph, and in such a way that the symbolic representation is capable of manipulation. We develop such a symbolism which allows, for example, the convolution of relationships. Thus we will see how the term first-cousin's first-cousin, when expressed symbolically, includes distinct simple relationships.

The most exhaustive discussion in this area comes, not surprisingly, from the anthropological literature. Atkins (1974*a,b*) discusses the representation of relationships between two individuals in terms of the two basic ones; viz, parent–offspring and husband–wife. Using these he develops methods of representing more complex relationships. We shall not go as deeply into the theory as Atkins, only developing the ideas sufficiently to give a clear picture of the scope of the methods, and to describe certain results of use later in the book. Similar ideas have been explored by, amongst others, Carnap (1958), Haldane and Jayakar (1962) and Kendall (1971).

We shall base our algebra, not on the two basic relationships between pairs of individuals, but rather on the two basic relationships possible between pairs of nodes in our graphical representation. Thus we are going to permit in our relationships not only

Figure 1.2(*c*). Types of inbreeding loop.

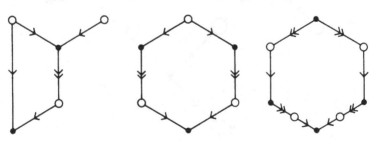

individuals but also marriages. We shall use as our basic symbols *R*, corresponding to a reproductive arc ↠, and *M*, corresponding to a marriage arc →. Thus if a marriage *b* has issue *B* we write *bRB*, and the symbol *R* on its own essentially denotes 'are the parents of'. If an individual *A* belongs to a marriage *a* we write *AMa*. We also employ the inverse notions, denoted by $\bar{R}$ and $\bar{M}$. Thus we write $B\bar{R}b$ to mean that *B* is the offspring of the marriage *b*, $\bar{R}$ alone denotes 'is the child of'. Similarly $a\bar{M}A$ means that the marriage *a* has as one of its members individual *A*.

We should emphasize at this point why we have chosen to work with *R* and *M*, rather than the parent-of (*P*), and mate-of (*M*) entities chosen by Atkins. The essential point is that it is couples who produce offspring, and not isolated individuals. The pairs of individuals are entities which should be included in any system of representation of relationships. We shall comment below on the contrast between our system and that of Atkins in particular cases.

A relationship between two individuals will thus be specified by a list of *R*s, *M*s, $\bar{R}$s and $\bar{M}$s corresponding to a sequence of arcs in the graph of the genealogy. Thus the sequence $\bar{R}\bar{M}MR$ corresponds to a simple sequence of arcs in the graph. However such a sequence of arcs, joining two individual nodes, does not uniquely define a relationship. Referring to Figure 1.3(*a*), suppose we trace the

Figure 1.3(*a*).

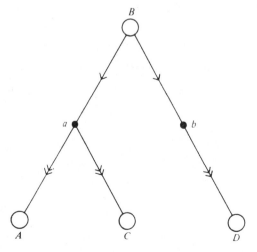

sequence $\bar{R}\bar{M}MR$ from A; now $\bar{R}\bar{M}$ takes us unambiguously to individual B, but now there is a choice of M since B is a member of two marriages. We may use M to reach marriage node b, and thence by R to individual D, or we may use M to reach marriage node a, and thence by R to either A or C. Thus the sequence $\bar{R}\bar{M}MR$ may represent the relationships, identity, sib or half-sib. We resolve this problem by introducing sequences which are constrained at each element not to retrace the previous step, and following Atkins we call these eversive. These relationship sequences will be written in parentheses, so that $(\bar{R}\bar{M}MR)$ is unambiguously half-sib.

It is a fairly simple matter to move from any sequence to the set of eversive sequences which it contains. Within any sequence there will exist neighbouring pairs, $\bar{R}R$, $M\bar{M}$, or $\bar{M}M$, which may be eversive, or non-eversive. Note that the pair $R\bar{R}$ is always non-eversive. We can obtain all the eversive sequences by considering all the $\bar{R}R$, $M\bar{M}$ and $\bar{M}M$ pairs, removing them if they are non-eversive, leaving them if they are eversive, and compiling a list of all the possible choices. As an example consider the sequence

Figure 1.3(*b*). Resolution of eversive (E) and non-eversive ($\bar{E}$) pairs.

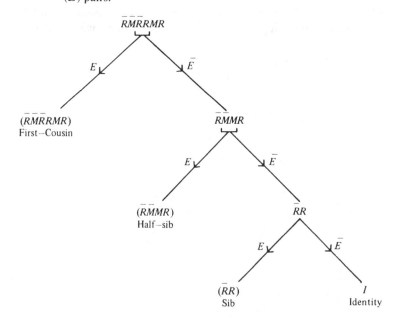

$\bar{R}\bar{M}\bar{R}RMR$. Here there is only one pair of interest, namely the $\bar{R}R$ pair. If this pair is taken as eversive then the whole relationship is eversive; if not then we reduce the relationship to $\bar{R}\bar{M}MR$ which now contains a pair $\bar{M}M$, which may or may not be eversive. The whole process of successive choices is set out in Figure 1.3(b). We see that the sequence in question contains four eversive relationships.

An added complication arises if there is more than one of the $\bar{R}R$, $M\bar{M}$ and $\bar{M}M$ pairs in the relationship sequence. We then have to make all possible removals, and we can do this by dealing with one pair at a time. The process is illustrated in Figure 1.3(c) for the sequence $\bar{R}\bar{M}R\bar{M}MRM\bar{M}$, and the various relationships shown in Figure 1.3(d) with reference to the individual A.

We can also use the above ideas to evaluate, for example, the relationship first-cousin's first-cousin, that is $(\bar{R}\bar{M}RRMR)$ $(\bar{R}\bar{M}\bar{R}RMR)$, where the $R\bar{R}$ in the centre of the combined sequence is necessarily non-eversive. At each stage we may only

Figure 1.3(c). Process of successive choices.

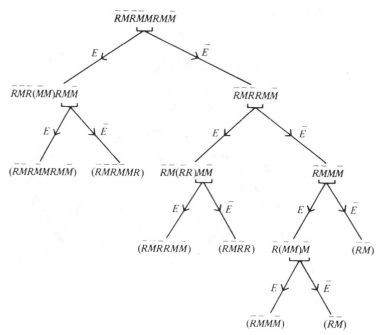

consider the central pair. The reduction of this relationship to its equivalent set of eversive relationships is particularly simple in this type of situation, and we have eversive relationships $(\bar{R}\bar{M}\bar{R}RM\bar{M}\bar{R}RMR)$, $(\bar{R}\bar{M}\bar{R}RMR)$, $(\bar{R}\bar{M}MR)$, $(\bar{R}R)$, I. The final four of these we have already identified as first-cousins, half-sib, sib and identity, the first being affinal.

We should note at this point that if one were to use as basic components the offspring–parent and spouse relationships, then a first-cousin, for example, is represented as $PP\bar{P}\bar{P}$, which is unambiguous only if no multiple marriage is allowed.

We list in Table 1.4(a) the sequences corresponding to some of the basic relationships. Note that for a sequence of length $2k$, there are $4 \times 3^{k-1}$ distinct relationships (though some of these may have the same term in English), these being all possible sequences in which $\bar{R}$ may be followed by $\bar{M}$ or R, M by $\bar{M}$ or R, $\bar{M}$ by $\bar{R}$ or M, but R by M only. The $4 \times 3^{k-1}$ sequences contain $2k+1$ biological relationships, so that the proportion of affinal relationships increases with k.

Figure 1.3(d).

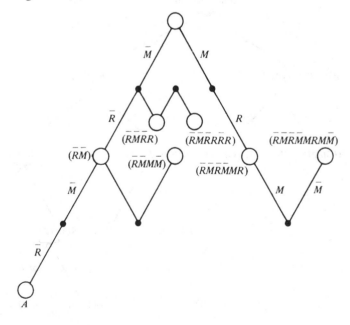

The incorporation of the sexes of some, or all, of the individuals concerned in the relationship can be accomplished easily. In any relationship certain of the implied arcs go from marriage node to individual, specifically those labelled $\bar{M}$ or R. If it is required to differentiate between the sexes of individuals, this may be done by using a superscript: $+$ for a male and $*$ for a female. Thus if we take the relation grandparent $(\bar{R}\bar{M}\bar{R}\bar{M})$, this can be differentiated into $(\bar{R}\bar{M}^+\bar{R}\bar{M}^+)$, $(\bar{R}\bar{M}^+\bar{R}\bar{M}^*)$, $(\bar{R}\bar{M}^*\bar{R}\bar{M}^+)$ and $(\bar{R}\bar{M}^*\bar{R}\bar{M}^*)$, corresponding to paternal grandfather, paternal grandmother, maternal grandfather and maternal grandmother. The manipulation of the sequences involving the sexes of individuals proceeds in basically the same way as those not involving sexes. For example in a sequence such as $\bar{M}^+\bar{R}R^*M$ the $\bar{R}R$ must be eversive as the sexes of the two individuals involved are different. The pairs which may be eversive or non-eversive are $^+\bar{R}R^+$, $^*\bar{R}R^*$, $\bar{M}^+M$ or $\bar{M}^*M$.

1.4 **Measures of relationship**

We have developed a method for representing relationships, and specified certain of these. In practice different societies deal with relationships in different ways, and may distinguish

Table 1.4(*a*)

English term	Sequence	Length	Biological
Parent	$(\bar{R}\bar{M})$		√
Sib	$(\bar{R}R)$	2	√
Spouse	$(M\bar{M})$		×
Offspring	(MR)		√
Grandparent	$(\bar{R}\bar{M}\bar{R}\bar{M})$		√
Uncle (Aunt)	$(\bar{R}\bar{M}\bar{R}R)$		√
Step-parent	$(\bar{R}\bar{M}M\bar{M})$		×
Half-sib	$(\bar{R}\bar{M}MR)$		√
Sib-in-law	$(\bar{R}RMM)$ or $(M\bar{M}\bar{R}R)$		×
Nephew (Niece)	$(\bar{R}RMR)$	4	√
Parent-in-law	$(M\bar{M}\bar{R}\bar{M})$		×
Spouse's spouse	$(M\bar{M}M\bar{M})$		×
Child-in-law	$(MRM\bar{M})$		×
Grandchild	$(MRMR)$		√
Step-child	$(M\bar{M}MR)$		×

between categories of relationship in different ways for different purposes. As we can see from Table 1.4(a) the English term sib-in-law is ambiguous being either a sib's spouse or a spouse's sib.

Atkins (1974a,b) has discussed at some length the derivation of certain measures of legal and social importance for biological relationships. We mention only two of these, by way of illustration, and how they are derived from our representation. Canon degree is derived simply by counting the number of individuals in the sequence from ego to the relative in question, and so in terms of our sequences is given by k. On the other hand, generational removal, which lumps individuals into generations, can be derived as the number of ($\bar{R}\bar{M}$)s in the sequence minus the number of (MR)s, the sign of the final result being dropped. This latter measure lumps an individual with his sibs, half-sibs, cousins etc., and puts parents, children, uncles, nephews together.

We shall see that measures of distance which are of importance for the geneticist need to take into account not only the relationship of two individuals, but also the genetic system. Thus a measure of relationship relevant for one character may be less relevant for some other character, if the modes of inheritance are different.

1.5   Genetic systems

Our main interest is in the application of the ideas of relationship to higher organisms, and in particular to man. Accordingly we confine our discussion to the genetic system usually found amongst those organisms, except where it is useful for purposes of illustration, or simplicity, to consider some alternative. We thus assume our system to be dioecious, and diploid.

We shall deal primarily with single loci, though the methods can be extended to linked loci, and will need to distinguish between the autosomal and sex-linked cases. In the former each individual receives a single gene from its father and a single gene from its mother, these being copies of randomly selected ones from the two parents. Thus the system is symmetric as far as the genetic component is concerned, and it does not require that we keep track of the sexes of individuals. On the other hand, the sex-linked case has a basic asymmetry between the sexes; a male has only one gene which was received from the mother, while a female has one from each parent. This requires that we keep track of the sexes of individual involved in a relationship, if our purpose involves genetic information.

Suppose, for example, that we are interested in the genetic relationship between two male first-cousins. The relationship first-cousins is specified by the string $(\bar{R}\bar{M}\bar{R}RMR)$, which has to be split into four strings if the sexes are to be specified; viz, (i) $(\bar{R}\bar{M}^+\bar{R}R^+MR)$, (ii) $(\bar{R}\bar{M}^*\bar{R}R^+MR)$, (iii) $(\bar{R}\bar{M}^+\bar{R}R^*MR)$ and (iv) $(\bar{R}\bar{M}^*\bar{R}R^*MR)$. For an autosomal locus there is no need to distinguish these four, but for a sex-linked locus the four sex-specific strings have markedly different consequences. For (i) both connecting relatives are male, so the cousins must have the same Y chromosome inherited from their common grandfather, but this carries no genetic information. For (ii) and (iii), where one connecting relative is male, the other female, there will be no genetic material derived from the same chromosome in a common grandparent. For (iv), where the connecting relatives are female, there is a probability that the X chromosomes are of common origin, and hence there is a genetic correlation. Thus we can see the importance of distinguishing such relationships in the sex-linked case, but not in the autosomal case.

When our focus is the population as a whole rather than a specific pedigree, additional causes of asymmetry may be present. Asymmetries of the marriage system, or of the demographic parameters, between the two sexes require that we take account of the sex of individuals in our analysis. Our measures of relationship will then need to specify the sexes of all the individuals involved in the relationship, including the individuals at the two ends.

2

Gene identity in genealogies

2.1 Identity by descent

The genetic consequences of a genealogical relationship arise from the fact that relatives may carry identical copies of genes received from their common ancestors. Such genes are said to be *identical by descent* (IBD), and in the absence of mutation such genes must be of the same allelic type. Thus relatives have an increased probability of carrying like genes, and this is reflected in observed correlations in phenotype between related individuals.

In any formal analysis of gene identity by descent, an origin must be defined. The set of originators of a population or *founders* may simply be the population existing at some point in time, although for some populations there may exist a natural set of founding individuals. Gene identity and genealogical relationships are specified only relative to this original population, whose genes are thus by definition distinct (non-IBD).

2.2 Kinship and inbreeding coefficients

For simplicity, we shall develop first the case of a pair of related individuals and consider a single autosomal locus, at which *Mendelian segregation* occurs. Thus every individual carries, at this locus, a maternal gene, which is an identical copy of a randomly chosen one of the two genes at this locus carried by his mother, and a paternal gene, which is an identical copy of a randomly chosen one of the two genes carried by his father.

A fundamental concept is that of the coefficient of *kinship*, or coefficient de *parenté*, Malecot (1948), also known as the *coancestry* (Falconer, 1960), *parentage* (Kempthorne, 1957) or *consanguinity* (Crow and Kimura, 1970), but distinct from Wright's (1922) coefficient of *relationship*. The coefficient of kinship, $\psi(B_1, B_2)$, between two individuals B_1 and B_2, is the probability that a gene chosen randomly from B_1 is IBD with a homologous gene chosen randomly from B_2. It follows, from the definition, that

$$\psi(B_1, B_2) = \psi(B_2, B_1), \qquad (2.2.1)$$

and from Mendelian segregation that

$$\psi(B, B) = \tfrac{1}{2}\{1 + \psi(F, M)\}, \qquad (2.2.2)$$

where individuals F and M are the parents of individual B.

Since related individuals may carry IBD genes, an individual whose parents are related may carry two homologous genes which are IBD. Such an individual is said to be *inbred*, and the coefficient of inbreeding, $\alpha(B)$, of an individual B, is the probability of this event. Thus, by Mendelian segregation,

$$\alpha(B) = \psi(F, M), \qquad (2.2.3)$$

where again F and M are the parents of B. Computation of these coefficients will be discussed later, in a more general context, but one useful result, which again follows directly from Mendelian segregation, is the following. If B_1 is neither a direct ancestor nor descendant of B_2

$$\psi(B_1, B_2) = \tfrac{1}{4}\{\psi(F_1, F_2) + \psi(F_1, M_2) + \psi(M_1, M_2) + \psi(M_1, F_2)\},$$
$$(2.2.4)$$

where F_i and M_i are the parents of B_i, $i = 1, 2$.

This result is of particular use in the analysis of population models, enabling us to write down recursions for the coefficients. For example, consider individuals who are sibs (S) at generation t, where we consider evolution relative to some origin $t = 0$. We have, from (2.2.4),

$$\psi_S^{(t)} = \tfrac{1}{4}\{2\psi^{(t-1)}(B, B) + 2\psi^{(t-1)}(B, B^*)\},$$

where the superscript denotes the generation, and B and B^* randomly chosen distinct individuals at the designated generation. Hence from (2.2.2) and (2.2.3) we have

$$\psi_S^{(t)} = \tfrac{1}{4}\{(1 + \alpha^{(t-1)}) + 2\psi^{(t-1)}\}$$
$$= \tfrac{1}{4} + \tfrac{1}{2}\psi^{(t-1)} + \tfrac{1}{4}\alpha^{(t-1)}, \qquad (2.2.5)$$

where the coefficients without arguments or subscripts denote the population values at the designated generation. Similarly for half-sibs (HS)

$$\psi_{HS}^{(t)} = \tfrac{1}{4}\{\tfrac{1}{2}(1 + \alpha^{(t-1)}) + 3\psi^{(t-1)}\}$$
$$= \tfrac{1}{8} + \tfrac{3}{4}\psi^{(t-1)} + \tfrac{1}{8}\alpha^{(t-1)}.$$ (2.2.6)

Where only relationships arising by virtue of segregation at the preceding generation are considered (see Chapter 3), these are the two main possibilities. Where overlapping generations are allowed, we have also the parent–offspring relationship (PO) (the offspring being at generation t). This provides again the recurrence (2.2.5). In some population models it is useful to include also the possibility of selfing. In this case there are three further relationships. First, there are sibs produced from a single parent by selfing (S*) which, by direct application of (2.2.4), gives

$$\psi_{S^*}^{(t)} = \psi^{(t-1)}(B, B) = \tfrac{1}{2}(1 + \alpha^{(t-1)}).$$ (2.2.7)

The parent–offspring relationship (PO*), the offspring of generation t being produced by selfing, gives the same result. Finally, we may have half-sibs, one of whom was produced by selfing of the common parent (HS*). Again (2.2.4) provides

$$\psi_{HS^*}^{(t)} = \tfrac{1}{4}\{2\psi^{(t-1)}(B, B^*) + 2\psi^{(t-1)}(B, B)\},$$

or again (2.2.5).

These equations will be used in Section 3.7, to provide recurrence equations for the population values of $H^{(t)} = 1 - \psi^{(t)}$ and $H_1^{(t)} = 1 - \alpha^{(t)}$.

Thus, for example, (2.2.5) will be used in the form

$$H_S^{(t)} = \tfrac{1}{2}H^{(t-1)} + \tfrac{1}{4}H_1^{(t-1)}.$$

For population models it is often more convenient to consider these probabilities of non-identity, rather than to use the coefficients of identity directly, but in the context of this chapter it is patterns of gene identity which are the basic element of the development.

2.3 Gene identity states: two individuals

For two individuals who are not inbred there are only three possible situations of gene identity with respect to their four genes at a single autosomal locus. Either each gene of each individual is IBD to one homologous gene of the other, or just one pair of genes (one in each individual) are IBD, or none of the four genes is IBD. Any genealogical relationship implies a probability for each of these three situations, and the phenotypic consequences of the relationship may be expressed in terms of these probabilities. Thus we have a space of genetic relationships

$$K = \{ \boldsymbol{k} = (k_2, k_1, k_0); k_2 + k_1 + k_0 = 1, k_i \geq 0 \}, \qquad (2.3.1)$$

where k_i = Prob(two individuals have i genes in common),

$$i = 0, 1, 2.$$

Each genealogical relationship implies a genetic one (for example, for *sibs*, $\boldsymbol{k} = (\frac{1}{4}, \frac{1}{2}, \frac{1}{4})$). Each genetic relationship implies a class of genealogical relationships (for example, $\boldsymbol{k} = (0, \frac{1}{2}, \frac{1}{2})$ corresponds to half-sibs, uncle–niece, grandparent–grandchild). The $\boldsymbol{k}$ values for a variety of genealogical relationships between non-inbred individuals B_1 and B_2 can be built up from the coefficients of kinship. As in Section 2.2 we shall denote by M_i and F_i the mother and father of $B_i (i = 1, 2)$. We have again (2.2.4);

$$\psi(B_1, B_2) = \tfrac{1}{4}\{\psi(F_1, F_2) + \psi(F_1, M_2) + \psi(M_1, M_2) + \psi(M_1, F_2)\}$$
$$(2.3.2)$$

provided B_1 is neither ancestor nor descendant of B_2, while in the absence of inbreeding (2.2.2) becomes

$$\psi(B, B) = \tfrac{1}{2} \qquad (2.3.3)$$

for any individual B. It is convenient to use also the less restrictive form of (2.3.2):

$$\psi(B_1, B_2) = \tfrac{1}{2}\{\psi(B_1, F_2) + \psi(B_1, M_2)\} \qquad (2.3.4)$$

provided B_2 is not a direct ancestor of B_1.

We note also

$$\psi(B_1, B_2) = \tfrac{1}{2}k_2(B_1, B_2) + \tfrac{1}{4}k_1(B_1, B_2), \qquad (2.3.5)$$

since if B_1 and B_2 have one [two] gene[s] in common the probability that random ones chosen from each are IBD is $\frac{1}{4}[\frac{1}{2}]$. Finally, van Aarde (1975) notes that

$$k_2(B_1, B_2) = \{\psi(M_1, M_2)\psi(F_1, F_2) + \psi(M_1, F_2)\psi(F_1, M_2)\},$$
$$(2.3.6)$$

since B_1 and B_2 have two genes in common if and only if they receive IBD genes from both same-sex parent pairs $((M_1, M_2)$ and $(F_1, F_2))$ or from both opposite-sex pairs $((M_1, F_2)$ and $(F_1, M_2))$.

For non-inbred offspring the pairs (M_1, F_1) and (M_2, F_2) have, of course, zero probability of contributing IBD genes.

We shall now consider some examples, but first, for convenience, derive the kinship coefficients for sibs (Figure 2.3(a)) and half-sibs (Figure (2.3(b)). In the former case we have, using (2.3.2),

$$\psi(B_1, B_2) = \tfrac{1}{4}\{\psi(F, F) + \psi(M, M) + 2\psi(M, F)\}$$
$$= \tfrac{1}{4} \qquad (2.3.7)$$

in the absence of inbreeding, from (2.3.3).

Figure 2.3. Examples of the derivation of probabilities k for non-inbred individuals. For further values see Figure 2.7 (a).

(a)   Sibs
$\psi = \tfrac{1}{4}$
$k = (\tfrac{1}{4}, \tfrac{1}{2}, \tfrac{1}{4})$

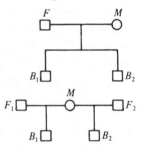

(b)   Half-sibs
$\psi = \tfrac{1}{8}$
$k = (0, 4\psi, 1 - 4\psi)$
$ = (0, \tfrac{1}{2}, \tfrac{1}{2})$

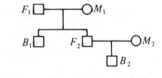

(c)   Uncle – niece
$\psi = \tfrac{1}{8}$
$k = (0, 4\psi, 1 - 4\psi)$
$ = (0, \tfrac{1}{2}, \tfrac{1}{2})$

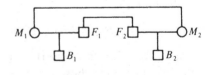

(d)   Parallel double-first-cousins
$\psi = \tfrac{1}{8}$
$k = (k_2, 4(\psi - \tfrac{1}{2}k_2), 1 - k_1 - k_2)$
$ = (\tfrac{1}{16}, \tfrac{3}{8}, \tfrac{9}{16})$

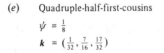

(e)   Quadruple-half-first-cousins
$\psi = \tfrac{1}{8}$
$k = (\tfrac{1}{32}, \tfrac{7}{16}, \tfrac{17}{32})$

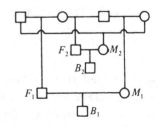

For half-sibs (Figure 2.3(b)) we have similarly

$$\psi(B_1, B_2) = \tfrac{1}{4}\{\psi(F_1, F_2) + \psi(M, M) + \psi(F_1, M) + \psi(F_2, M)\}$$
$$= \tfrac{1}{8}, \qquad (2.3.8)$$

in the absence of inbreeding, provided F_1 and F_2 are unrelated.

Consider now a unilateral relationship, such as uncle–niece (Figure 2.3(c)); that is, only one parent of B_2 is related to B_1 (or vice versa). For such a relationship $k_2(B_1, B_2) = 0$, since both terms of (2.3.6) contain a zero factor. From (2.3.4) we have

$$\psi(B_1, B_2) = \tfrac{1}{2}\psi(B_1, F_2) \qquad (\psi(B_1, M_2) = 0; \text{Figure 2.3}(c))$$
$$= \tfrac{1}{8} \quad \text{from (2.3.7).}$$

Thus

$$k_1(B_1, B_2) = 4\psi(B_1, B_2) = \tfrac{1}{2} \quad \text{from (2.3.5),}$$

or

$$(k_2, k_1, k_0) = (0, \tfrac{1}{2}, \tfrac{1}{2}).$$

As an example of a bilateral relationship we consider (parallel) double-first-cousins (Figure 2.3(d)). From (2.3.6) and (2.3.7)

$$k_2(B_1, B_2) = \psi(M_1, M_2)\psi(F_1, F_2) = \tfrac{1}{16}.$$

From (2.3.2) and (2.3.7)

$$\psi(B_1, B_2) = \tfrac{1}{4}(\tfrac{1}{4} + 0 + \tfrac{1}{4} + 0) = \tfrac{1}{8}.$$

Hence from (2.3.5)

$$k_1(B_1, B_2) = 4(\psi - \tfrac{1}{2}k_2) = \tfrac{3}{8}$$

and

$$\boldsymbol{k} = (k_2, k_1, k_0) = (\tfrac{1}{16}, \tfrac{3}{8}, \tfrac{9}{16}).$$

Most bilateral relationships not involving inbreeding are as this one, in that either $\psi(F_1, F_2) = \psi(M_1, M_2) = 0$ or $\psi(F_1, M_2) = \psi(M_1, F_2) = 0$. That is, either the parents of the same sex are related (parallel relationships), or those of opposite sex are related (cross-relationships), but not both. However, it is possible to have all four parental kinships of (2.3.6) non-zero, without inbreeding of offspring (Trustrum, 1961). The simplest example is quadruple-half-first-cousins (Figure 2.3(e)). Here all cross-parental pairs are half-sibs and thus, using (2.3.8),

$$k_2(B_1, B_2) = \tfrac{1}{8} \times \tfrac{1}{8} + \tfrac{1}{8} \times \tfrac{1}{8} = \tfrac{1}{32} \quad \text{from (2.3.6)}$$

and

$$\psi(B_1, B_2) = \tfrac{1}{4}(\tfrac{1}{8} + \tfrac{1}{8} + \tfrac{1}{8} + \tfrac{1}{8}) = \tfrac{1}{8} \quad \text{from (2.3.2).}$$

Hence
$$k_1(B_1, B_2) = 4(\psi - \tfrac{1}{2}k_2) = \tfrac{7}{16} \quad \text{from (2.3.5)}$$
and
$$k = (k_2, k_1, k_0) = (k_2, k_1, 1 - k_1 - k_2) = (\tfrac{1}{32}, \tfrac{7}{16}, \tfrac{17}{32}).$$

Note that kinship coefficients are insufficient to characterize a relationship; (*b*), (*d*) and (*e*) all have $\psi = \tfrac{1}{8}$, but the probabilities k which determine genotype distributions are different. A table of gene identity state probabilities is given in Section 2.7, where a diagrammatic representation of the space (2.3.1) is given. Further values of k are given by Jacquard (1974, p. 113) and Thompson (1975*a*). These may all be verified by the method of the above examples.

The parameters k_i are (essentially) the Cotterman k-coefficients. Cotterman (1940) developed their use in obtaining expressions for joint and conditional phenotypic distributions, and hence in risk analysis. Li and Sacks (1954) also use these three states (in their notation providing matrices I, T and O) in their recursive matrix derivations of pairwise phenotype distributions, via the fundamental relationships of 'sibs' and 'parent–offspring'.

More generally, for individuals who may be inbred, there are fifteen possible states of gene identity, provided the maternal and paternal genes of the two individuals are distinguished. These were specified by Gillois (1965), and are shown diagrammatically in Table 2.3(*a*). Again any genealogical relationship implies a probability distribution over these fifteen states, which in turn determines the probability of any consequential genotypic event. However, some identity states are genotypically indistinguishable. Since the genotype of an individual is the *unordered* pair of allelic types of his two genes, the fifteen states fall into nine classes (Table 2.3(*a*)). For example, the four states in which either the maternal or the paternal gene (but not both) of the first individual B_1 is IBD to either the maternal or the paternal gene (but not both) of the second individual B_2 are all equivalent. We see, in particular, that the seven states in which neither individual has two IBD genes reduce to the three classes described above of two, one or zero genes common to the two individuals.

Rather than the fifteen probabilities of the individual states, only the nine probabilities of the genetically distinct situations of gene

Table 2.3(a). *The identity states between two individuals and their partition*

Elementary state	Gillois (1965)	Thompson (1974a)	Identity under transformations*	Equivalence under transformation*	Condensed State (Jacquard, 1972)	Orbit Labelling (Thompson, 1974a)	Probability
1	□	(1, 1, 1, 1)	T_1, T_2	—	$S_1 =$	(1, 1, 1, 1)	Π_1
2	⊔	(1, 1, 1, 2)	T_1 ⎫				
3	⊓	(1, 1, 2, 1)	T_1 ⎬	T_2	$S_3 =$	(1, 1, 1, 2)	Π_3
4	⊔	(1, 2, 1, 1)	T_2 ⎫				
5	⊓	(1, 2, 2, 2)	T_2 ⎬	T_1	$S_5 =$	(1, 2, 1, 1)	Π_5
6	‖	(1, 1, 2, 2)	T_1, T_2	—	$S_2 =$	(1, 1, 2, 2)	Π_2
7	::	(1, 1, 2, 3)	T_1, T_2	—	$S_4 =$	(1, 1, 2, 3)	Π_4
8	⊏	(1, 2, 3, 3)	T_1, T_2	—	$S_6 =$	(1, 2, 3, 3)	Π_6
9	X	(1, 2, 1, 2)	T_1 ⎫				
12	⋮	(1, 2, 2, 1)	T_1 ⎬	T_2	$S_7 =$	(1, 2, 1, 2)	Π_7
10	∕	(1, 2, 1, 3)	—				
11	⋅∖⋅	(1, 2, 2, 3)	—				
13	⋅⋅	(1, 2, 3, 1)	—	T_1, T_2	$S_8 =$	(1, 2, 1, 3)	Π_8
14	::	(1, 2, 3, 2)	—				
15	⋅⋅	(1, 2, 3, 4)	T_1, T_2	—	$S_9 =$	(1, 2, 3, 4)	Π_9

* see text

identity are required to determine pairwise genotype distributions. Conversely only the nine total probabilities of these situations are identifiable parameters of pairwise genotype distributions (see Section 2.5). These nine classes of states have been called the 'genetically distinct' gene identity states by Jacquard (1972). We note that Cotterman (1940) distinguished twelve states, since he differentiated between the maternal and paternal gene of one individual but not the other.

2.4 Gene identity: any number of individuals

The theory of gene identity states, equivalence classes of states (i.e. *genetically distinct* states), and the derivation of joint or conditional phenotypic distributions in terms of the gene-identity-state probabilities, have been extended from two individuals to (in theory) an arbitrary number (Thompson, 1974a). A gene identity state may be specified via the following algorithm:

(*a*) Order the *n* individuals and the two genes within each. A state will be designated by a sequence of gene labels (integers):

$$S = (g_1, \ldots, g_{2i-1}, g_{2i}, g_{2i+1}, \ldots, g_{2n}),$$

where g_{2i-1} and g_{2i} are the labels assigned to the maternal and paternal gene (respectively) of the *i*th individual.

(*b*) $g_1 = 1$.

(*c*) Suppose $g_1, \ldots, g_{j-1}$ have been assigned values in the set $\{1, \ldots, k\}$ $(k \leq j - 1)$; if the *j*th gene is IBD to any preceding gene (say the *l*th) $g_j = g_l$; if not $g_l = k + 1$.

This procedure, equivalent to that of Nadot and Vaysseix (1973), enables recurrence relations for the numbers of gene identity states between $2n$ genes to be derived. These have been given by Thompson (1974a), and the results tabulated. The total number of states increases rapidly with n; for $n = 6$ there are 4 213 597 states. However these fall into only 198 091 genetically distinct classes of states. With the above specification of states, the genetically distinct classes of states may be readily determined, since these are the equivalence classes of orbits under the following group of transformations.

Let H_n be the group generated by

$$\{T_i, 1 \leq i \leq n\},$$

where

$$T_i(g_1, \ldots, g_{2i-1}, g_{2i}, g_{2i+1}, \ldots, g_{2n})$$

$$= (g_1, \ldots, g_{2i-2}, g_{2i}, g_{2i-1}, g_{2i+1}, \ldots, g_{2n}).$$

That is T_i interchanges the maternal and paternal genes of the ith individual. Since the equivalence of gene identity states is a consequence of the genotypic indistinguishability of the maternal and paternal genes of an individual, it is immediate that the classes of equivalent states are the orbits under H_n. This enables the numbers and sizes of equivalence classes, and numbers comprising different numbers of distinct genes, to be investigated. Further details are given by Thompson (1974a).

A practical problem is the computation of gene-identity-state probabilities for a specified genealogical relationship. Although Wright (1922) resolved the question of computation of pairwise kinship coefficients, the general problem remains unsolved. There are many algorithms for the computation of pairwise kinship and inbreeding coefficients (see, for example, Stevens (1975)), but only that of Nadot and Vaysseix (1973) provides the fifteen probabilities of gene-identity-states between an arbitrary set of up to four genes in a pedigree. An extension of their algorithm allows (in theory) the computation of gene-identity-state probabilities and distinct-gene-identity-state probabilities for any genealogical relationship (Thompson, 1974b, 1980a), but practical problems arise with the very large numbers of possible states.

2.5   Joint and conditional genotype probabilities

The major importance of the specification of gene identity states, and computation of their probabilities, is in the derivation of joint or conditional phenotype probabilities. These in turn are of use in genetic counselling and other practical problems. We consider first the case of genotype probabilities, at a single autosomal locus with alleles A_i, $1 \leq i \leq r$, of known population frequencies. Jacquard (1972) has considered conditional probabilities for the genotype of an individual, conditional on that of a relative. (In fact, his result contains a slight error, which has been corrected by Elston and Lange (1976).) However, in general, it is more convenient to consider joint probabilities; any conditional probability may clearly be obtained as the ratio of the two relevant joint ones. Thompson (1974a) provides the following representation of joint probabilities.

Suppose we have n individuals ($n \geq 1$) in a genealogical relationship R implying probabilities $\{\Pi_i : 1 \leq i \leq D_n\}$ over the set of D_n distinct-gene-identity-states $\{S_i : 1 \leq i \leq D_n\}$. Let the genotypes of the n ordered individuals be denoted by $\boldsymbol{G} = (G_1, \ldots, G_n)$. Then

$$\text{Prob }(\boldsymbol{G}|R) = \sum_{i=1}^{D_n} \Pi_i \text{ Prob }(\boldsymbol{G}|S_i). \qquad (2.5.1)$$

Now for a particular state S, in which there are $k(S)$ distinct genes, we may write

$$\text{Prob }(\boldsymbol{G}|S) = \sum_{\mathscr{A} \in \Omega(\boldsymbol{G}:S)} \left\{ \prod_{j=1}^{k(S)} q_j(\mathscr{A}) \right\}, \qquad (2.5.2)$$

where $\Omega(\boldsymbol{G}:S)$ is the set of all allocations $\mathscr{A}$ of allelic types to the $k(S)$ labelled genes which result in ordered genotypes $\boldsymbol{G}$ under S, and $q_j(\mathscr{A})$ is the population frequency of the allelic type assigned to the gene labelled j under allocation $\mathscr{A}$.

For example, consider the (class of states equivalent to) state $S = (1, 2, 1, 3, 1, 4)$ between three individuals, each of whom has genotype $A_1 A_2$. There are two possible allocations:

$$\mathscr{A}_1; q_1(\mathscr{A}_1) = p_1, \qquad q_2(\mathscr{A}_1) = q_3(\mathscr{A}_1) = q_4(\mathscr{A}_1) = p_2$$

$$\mathscr{A}_2; q_1(\mathscr{A}_2) = p_2, \qquad q_2(\mathscr{A}_2) = q_3(\mathscr{A}_2) = q_4(\mathscr{A}_2) = p_1,$$

where p_1 and p_2 are the population allele frequences of A_1 and A_2.

Thus $\text{Prob }(\boldsymbol{G}|S) = p_1 p_2^3 + p_1^3 p_2$, for these genotypes under this particular identity state.

Combining equations (2.5.1) and (2.5.2)

$$\text{Prob }(\boldsymbol{G}|R) = \sum_{i=1}^{D_n} \Pi_i \left\{ \sum_{\mathscr{A} \in \Omega(\boldsymbol{G}:S_i)} \left\{ \prod_{j=1}^{k(S_i)} q_j(\mathscr{A}) \right\} \right\}. \qquad (2.5.3)$$

Although this equation appears complicated, simple algorithms to determine the sets $\Omega(\boldsymbol{G}:S)$ may be developed. The joint probability of a set of phenotypes, $\boldsymbol{\phi} = (\phi_1, \ldots, \phi_n)$, determined by the single autosomal locus, may then be written as

$$\text{Prob }(\boldsymbol{\phi}|R) = \sum_{\boldsymbol{G}} \text{Prob }(\boldsymbol{\phi}|\boldsymbol{G}) \text{ Prob }(\boldsymbol{G}|R)$$

$$= \sum_{\substack{G_i \in C(\phi_i) \\ 1 \leq i \leq n}} \cdots \sum \left\{ \prod_{i=1}^{n} \text{Prob }(\phi_i|G_i) \right\} \text{Prob }(\boldsymbol{G}|R), \qquad (2.5.4)$$

where $C(\phi_i)$ is the set of genotypes which can give rise to phenotype ϕ_i. The probabilities Prob $(\phi_i|G_i)$, specifying the genotype–phenotype relationship, are a basic element in the probability distributions of phenotypes on pedigrees; we shall call them *penetrance* probabilities. Although equations (2.5.1) to (2.5.4) imply lengthy computations when n is more than four, algorithms may be implemented to perform these efficiently, (Thompson, 1980a). An example of the use of Sections 2.4 and 2.5 to derive joint phenotype distributions is given in Section 6.4.

2.6   Phenotypic similarities between relatives

In practice, it is the phenotypic similarities between relatives that are observed, and conditional phenotype probability distributions that are required for risk analysis. In general, for phenotypes controlled by a single locus, (2.5.4) gives the required probabilities, or probability densities. However, it is also illuminating to consider in more detail the correlation between the phenotypes of a pair of relatives for a quantitative character controlled by such a locus.

Following Jacquard (1974) we consider an r-allele autosomal locus, with alleles $A_i(1 \leq i \leq r)$ having frequencies $p_i(1 \leq i \leq r)$. We assume a quantitative character for which the mean of the genotypic component, for an individual of genotype A_iA_j, is

$$\mu_{ij} = \mu + a_i + a_j + d_{ij}. \tag{2.6.1}$$

Here a_i are the additive effects of alleles A_i, and d_{ij} the dominance deviations (Fisher, 1930). Thus, under Hardy–Weinberg equilibrium, the population mean is

$$\mu = \sum_i \sum_j p_i p_j \mu_{ij}$$

and

$$a_i = \sum_j p_j(\mu_{ij} - \mu). \tag{2.6.2}$$

Thus

$$\left. \begin{array}{l} \sum_i p_i a_i = 0 \\[2mm] \sum_i p_i d_{ij} = \sum_i p_i(\mu_{ij} - a_i - a_j - \mu) = 0 \end{array} \right\} \forall j \tag{2.6.3}$$

and

$$\sum_i \sum_j p_i p_j a_i d_{ij} = 0.$$

Table 2.6(a). *The mean values of genetic contributions to a quantitative character*

Class of states	Probabilities given R	B_1 genotype	B_1 probabilities	B_1 mean	B_2 mean, given B_1 and state
$(1,1,1,1)$	Π_1	A_iA_i	p_i	$\mu+2a_i+d_{ii}$	$\mu+2a_i+d_{ii}$
$(1,1,2,2)$	Π_2	A_iA_i			$\sum p_i(\mu+2a_i+d_{ii})=\mu+E_{\mathrm{H}}(d)$
$(1,1,1,2)$	Π_3	A_iA_i			$\sum p_i(\mu+a_i+a_i+d_{ii})=\mu+a_i$
$(1,1,2,3)$	Π_4	A_iA_i			$\sum\sum p_ip_j(\mu+a_i+a_j+d_{ij})=\mu$
$(1,2,1,1)$	Π_5	A_iA_j			$\frac{1}{2}(\mu+2a_i+d_{ii})+\frac{1}{2}(\mu+2a_j+d_{jj})$ $=\mu+a_i+a_j+\frac{1}{2}(d_{ii}+d_{jj})$
$(1,2,3,3)$	Π_6	A_iA_j	$(2-\delta_{ij})^*p_ip_j$	$\mu+a_i+a_j+d_{ij}$	$\sum p_i(\mu+2a_i+d_{ii})=\mu+E_{\mathrm{H}}(d)$
$(1,2,1,2)$	Π_7	A_iA_j			$\mu+a_i+a_j+d_{ij}$
$(1,2,1,3)$	Π_8	A_iA_j			$\mu+\sum p_i(\frac{1}{2}(\mu+a_i+a_i+d_{ii})+\frac{1}{2}(\mu+a_i+a_l+d_{il}))$ $=\mu+\frac{1}{2}(a_i+a_l)$
$(1,2,3,4)$	Π_9	A_iA_j			μ

* $\delta_{ij}=1$ if $i=j$; 0 otherwise

Now consider two individuals B_1 and B_2, with genetic components Y_1 and Y_2 for the quantitative character. The individuals have their genes in one of the nine identity states S_i of Figure 2.3(a), with probabilities Π_i given by their genealogical relationship R. These states are given again in Table 2.6(a), together with joint mean values of the individuals' phenotypes. For example, if B_1 has two identical genes he has genotype A_iA_i with probability p_i, and mean value $\mu + a_i + d_{ii}$. Under identity state $(1, 1, 1, 2)$ (in the notation of Section 2.4) B_2 is of genotype A_iA_j with probability p_j, and thus has mean

$$\sum_j p_j(\mu + a_i + d_j + d_{ij}) = \mu + a_i.$$

For the final five states B_1 has two non-identical genes and is of genotype A_iA_j with probability $(2 - \delta_{ij})p_ip_j$ (where δ_{ij} is the Kronecker delta). Under state $(1, 2, 1, 3)$, for example, B_2 is of genotype A_iA_l or A_jA_l each with probability $\frac{1}{2}p_l$, and hence has mean value

$$\mu + \tfrac{1}{2}(a_i + a_j).$$

Thus we may compute the expected value of $Y_1 Y_2$:

$$E(Y_1 Y_2 | R) = \sum_{i=1}^{9} \left[\Pi_i \right.$$
$$\left. \times \sum_g \{ P(g(B_1) = g | S_i) \mu(B_1 | g) \mu(B_2 | g(B_1) = g, S_i) \} \right],$$

where the second summation is over genotypes g of B_1, $\mu(B_1 | g)$ is the mean for an individual of genotype g and $\mu(B_2 | g(B_1) = g, S_i)$ is the mean for relative B_2 given the genotype g of B_1, and gene identity state S_i.

We may also compute the overall means

$$E(Y_1) = \sum_{i=1}^{9} \Pi_i \sum_g p(g(B_1) | S_i) \mu(B_1 | g(B_1) = g)$$
$$= \left\{ \sum_{1}^{4} \Pi_i \right\} \{ \sum p_i(\mu + 2a_i + d_{ii}) \}$$
$$+ \left\{ \sum_{5}^{9} \Pi_i \right\} \{ \sum\sum p_ip_j(\mu + a_i + a_j + d_{ij}) \}$$
$$= \mu + (\Pi_1 + \Pi_2 + \Pi_3 + \Pi_4) \sum p_i d_{ii}$$
$$= \mu + f_1 E_H(d),$$

where f_1 is the inbreeding coefficient of B_1 and $E_H(d) = \sum p_i d_{ii}$ is the

expected dominance deviation in a fully inbred line. The subscript
H will denote such a line. Similarly $E(Y_2) = \mu + f_2 E_H(d)$ and hence

$$\text{cov}(Y_1, Y_2) = E(Y_1 Y_2) - (EY_1)(EY_2)$$

$$= \{4\Pi_1 + 2(\Pi_3 + \Pi_5 + \Pi_7) + \Pi_8\} \sum_i p_i a_i^2$$

$$+ (4\Pi_1 + \Pi_3 + \Pi_5) \sum_i p_i a_i d_{ii} + \Pi_1 \sum p_i d_{ii}^2$$

$$+ \Pi_7 \sum \sum p_i p_j d_{ij}^2 + (D_2 - f_1 f_2)(E_H(d))^2 \qquad (2.6.4)$$

Now $2 \sum p_i a_i^2$ is the additive genetic variance, V_A, and $\sum \sum p_i p_j d_{ij}^2$ is
the dominance variance, V_D, (Fisher, 1930). We may also write
$E_H(d) = \sum p_i d_{ii}$ for the mean dominance effect in a fully inbred
population (see above), and $V_H(d) = \sum p_i d_{ii}^2 - (E_H(d))^2$ for its variance, while $\text{cov}_H(a, d) = \sum p_i (2a_i) d_{ii}$ is the covariance between
additive and dominance effects in such an inbred line. Also $4\Pi_1 +
2(\Pi_3 + \Pi_5 + \Pi_7) + \Pi_8 = 4\psi(B_1, B_2)$ and $4\Pi_1 + \Pi_3 + \Pi_5 = 4Q(B_1, B_2)$,
where $Q(B_1, B_2)$ is the probability that a set of three genes,
randomly chosen from the four of B_1 and B_2, are all identical by
descent. Further $\Pi_1 + \Pi_2 - f_1 f_2$ is the *covariance in inbreeding*
between B_1 and B_2, $(I(B_1, B_2))$ since $\Pi_1 + \Pi_2$ is the probability both
individuals carry identical genes. Thus

$$\text{cov}(Y_1, Y_2) = 2\psi(B_1, B_2)V_A + \Pi_7 V_D + \Pi_1 V_H(d)$$

$$+ 2Q(B_1, B_2)\text{cov}_H(a, d)$$

$$+ I(B_1, B_2)(E_H(d))^2. \qquad (2.6.5)$$

Formula (2.6.5) is essentially that of Harris (1964), Gillois (1964)
and Jacquard (1974), although it differs from the last mentioned
owing to some slight errors in that derivation. The first two terms
are the only ones which occur for non-inbred individuals, and these
are of course well-known (see, for example, Crow and Kimura
(1970, p. 139). The extension to inbred relatives may however be of
some importance where data are available on an extensive pedigree
of highly inbred individuals, such as that of the population of Tristan
da Cunha (Roberts, 1971). However it must be noted that (2.6.5) is
only a covariance in *genetic* effects; common environment and
gene–environment interactions may generate further similarity
between relatives.

Phenotypic similarities between relatives may be viewed predic-
tively or inferentially. So far we have considered only the joint

phenotype distributions for individuals in a given genealogical relationship. The inference problem is to consider what information about the genealogical relationship is provided by an observed set of joint phenotypes. This has been considered by Thompson (1975a). The joint phenotype distribution for a character determined by a single autosomal locus is given by (2.5.4). This is a linear function of the joint-gene-identity-state probabilities, and is the likelihood for these parameters when $\{\phi_i; 1 \leq i \leq n\}$ is a given set of observed phenotypes. When several characters, controlled by unlinked autosomal loci, are observed, the likelihood function is the product of such linear functions.

Estimation of the parameters of this likelihood function will be discussed in Chapter 5, but we consider here a problem raised by this estimation. Given maximum likelihood gene-identity-state probabilities, to what genealogical relationship does this correspond? We can compare the likelihoods of alternative specified relationships, but this simply evades the issue. In practice a *network* is required of genealogical relationships spanning the space of identity-state probabilities. For more than three individuals this is a massive task. Furthermore, in investigating the cover of the identity-state-probability space, another problem is encountered. The space of probabilities $\{\Pi_i; \Pi_i \geq 0, \sum_1^{D_n} \Pi_i = 1\}$ is *not* covered by the space of genealogical relationships between n individuals. By this we mean not merely that identity-state probabilities must be dyadic rationals and can thus only be dense in the space, but that there are open regions of the space which contain no genealogical relationships. This is perhaps the most interesting open question: what is the true space of attainable gene-identity-state probabilities?

2.7 The attainable gene identity space

To illustrate this problem, we consider first the case of two non-inbred relatives and the space $K = \{(k_2, k_1, k_0); k_i \geq 0, k_2 + k_1 + k_0 = 1\}$ of (2.3.1). Figure 2.7(a) shows this space, with various standard genealogical relationships marked upon it. Now, as in Section 2.3, for any two such individuals B_1 and B_2 we have

$$\text{and} \quad \left. \begin{array}{l} k_2(B_1, B_2) = \psi_{\mathrm{mm}}\psi_{\mathrm{ff}} + \psi_{\mathrm{mf}}\psi_{\mathrm{fm}} \\ \psi(B_1, B_2) = \tfrac{1}{4}(\psi_{\mathrm{mm}} + \psi_{\mathrm{ff}} + \psi_{\mathrm{mf}} + \psi_{\mathrm{fm}}), \end{array} \right\} \quad (2.7.1)$$

provided neither individual is an ancestor of the other (see (2.3.2) and (2.3.6)), where ψ_{mm} etc. are the four cross-parental kinship coefficients. It is possible for all four of these coefficients to be non-zero, or for each of the father and the mother of each individual

Figure 2.7(*a*). Some standard relationships; *k*-coefficients and triangle representation.

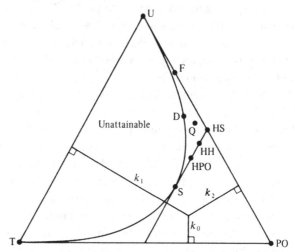

Key	Relationship	k_2	k_1	k_0	ψ
U	Unrelated	0	0	1	0
T	Twin	1	0	0	$\frac{1}{2}$
PO	Parent–offspring	0	1	0	$\frac{1}{4}$
S	Sib	$\frac{1}{4}$	$\frac{1}{2}$	$\frac{1}{4}$	$\frac{1}{4}$
HS	Half-sib	0	$\frac{1}{2}$	$\frac{1}{2}$	$\frac{1}{8}$
Q	Quadruple-half-first-cousin	$\frac{1}{32}$	$\frac{14}{32}$	$\frac{17}{32}$	$\frac{1}{8}$
D	Double-first-cousins	$\frac{1}{16}$	$\frac{6}{16}$	$\frac{9}{16}$	$\frac{1}{8}$
F	First-cousins	0	$\frac{1}{4}$	$\frac{3}{4}$	$\frac{1}{16}$
HH	Half-sibs whose non-identical parents are half-sibs	$\frac{1}{16}$	$\frac{1}{2}$	$\frac{7}{16}$	$\frac{5}{32}$
HPO	Half-sibs whose non-identical parents are parent–offspring or sibs	$\frac{1}{8}$	$\frac{1}{2}$	$\frac{3}{8}$	$\frac{3}{16}$

to be related to both the father and the mother of the other, without either B_1 or B_2 being inbred (see Figure 2.3(e)). The restriction that (2.7.1) holds only where B_1 is neither ancestor nor descendant of B_2 is not relevant to the current problem. Where this is the case, we must have $k_2(B_1, B_2) = 0$ if neither individual is inbred, and thus the space is

$$K = \{(0, k_1, k_0); k_1 + k_0 = 1\},$$

the whole of which may be attained. (As already described we speak of attaining the whole of a space, where points dense in the space can be attained.)

Now

$$\psi(B_i, B_2) = \tfrac{1}{4}\{2k_2(B_1, B_2) + k_1(B_1, B_2)\}. \tag{2.7.2}$$

As shown by Thompson (1976b), two applications to (2.7.1) of the geometric mean–arithmetic mean inequality provide

$$k_2(B_1, B_2) \le 4(\psi(B_1, B_2))^2$$

and hence, using (2.7.2),

$$4k_2(B_1, B_2)k_0(B_1, B_2) \le (k_1(B_1, B_2))^2. \tag{2.7.3}$$

Thus only a small part of the space K can be attained by any genealogical relationship; two-thirds of the region is excluded by the restriction (2.7.3) (see Figure 2.7(a)).

We note also that there are points corresponding to true genealogical relationships arbitrarily close to any k satisfying (2.7.3). If we put

$$\psi_{mf} = \psi_{fm} = 0$$

(2.7.1) becomes

$$k_2 = \psi_{mm}\psi_{ff}$$

and

$$k_1 = \psi_{mm}(1 - \psi_{ff}) + \psi_{ff}(1 - \psi_{mm}).$$

For given k these can be solved for ψ_{mm} and ψ_{ff}; the quadratic equation for ψ has real roots lying between 0 and 1 if and only if k satisfies (2.7.3). Since we may construct a relationship providing any such dyadic rational kinship coefficient, provided we allow consideration of inbred individuals, we may obtain a pair of fathers and a pair of mothers, the fathers being unrelated to the mothers. The non-inbred offspring of the couples formed from these individuals will have the required relationship k. It is also of interest to characterize the *extreme* relationships, giving equality in (2.7.3). These are relationships where the coefficient of kinship between the mothers of B_1 and B_2 is equal to the coefficient between the fathers, the mothers being unrelated to the fathers, for example sibs and double-first-cousins (Figure 2.7(a)).

The above restriction on the attainable gene-identity-state probabilities is simply a result of Mendelian segregation. It is clear that similar restrictions will hold in the spaces of coefficients for any number, n, of individuals, and in general only a very small part of the space

$$\{\{\Pi_i : 1 \le i \le D_n\}; \Pi_i \ge 0, \sum_1^{D_n} \Pi_i = 1\}$$

can be attained. However even for two individuals the problem is not fully solved. Thompson (1978) has given some simple restrictions on the eight-dimensional space of the nine distinct-gene-identity-state probabilities, and Thompson (1980b) has considered in detail the possible combinations of probabilities for an individual, B_1, and a direct descendant, B_2, who is not also a collateral relative. In this case identity between the genes of B_1 is independent of any identity between B_1 and B_2 and the problem reduces to one of the five parameters. These are most readily characterized as

$P = \text{Prob } (B_2 \text{ receives both genes from } B_1)$
$R = \text{Prob } (B_2 \text{ receives no genes from } B_1)$ $\Big\}\, Q = (1 - P - R)$

$f = \text{Prob } (B_1 \text{ has IBD genes})$

$f^* = \text{Prob } (B_2 \text{ has IBD genes } not \text{ from } B_1)$

and

$\lambda = \text{Prob } (B_2 \text{ has IBD genes} \,|\, \text{both genes of } B_2 \text{ derive from } B_1).$

The expression of the nine identity state probabilities in terms of these parameters provides some restrictions on the identity-state space.

For example,

$$\begin{aligned}
D(1122)D(1234) &= \{ff^*R\}\{(1-f)(1-f^*)R\} \\
&= f(1-f^*)Rf^*(1-f)R \\
&= D(1123)D(1233),
\end{aligned}$$

(2.7.4)

where the states of the ordered pair (B_1, B_2) are labelled as in Section 2.3. It is also readily seen that $\lambda \geq \frac{1}{2}$, which gives $D(1211) \geq D(1212)$. Other simple restrictions of this kind are given by Thompson (1980*b*).

There is however a more interesting class of restrictions. The combination

$$Q^2 - 4PR = 1 - 2(R + P) + (P - R)^2$$

(cf. (2.7.3)) is a measure of the correlation in descent from B_1 between the maternal and paternal gene of B_2 imposed by any common paths of descent from B_1 to intermediate bilateral ancestors of B_2. The parameter λ is also a measure of this correlation, but not an equivalent one. With regard to overall correlation in descent, the number of offspring of B_1 through whom paths pass is irrelevant, since all carry precisely one of B_1's genes. With regard to the parameter λ, which measures correlation in descent from a single gene of B_1 rather than from B_1 as an individual, paths through the same offspring individual necessarily transmit the same B_1-gene. To investigate this problem further it is convenient to introduce further parameters

$$\gamma_m = \text{Prob (paternal gene of } B_2 \text{ derives from } B_1)$$

$$\gamma_f = \text{Prob (maternal gene of } B_2 \text{ derives from } B_1)$$

and

$$h = \{\gamma_m(1 - \gamma_m)\gamma_f(1 - \gamma_f)\}^{1/2}.$$

We may then express correlation in descent more directly as

$$\rho = (P - \gamma_m\gamma_f)/h$$

and derive, exactly as in the derivation of (2.7.3), the restriction

$$Q^2 - 4PR \geq -\rho h. \tag{2.7.5}$$

The question thus finally becomes one of the relationship between ρ and λ. It is clear that $\lambda = \frac{1}{2}$ implies $\rho = 0$, and $\rho = 1$ implies $\lambda = 1$. It also seems that for any given ρ there is a smallest attainable λ, and for any given λ a largest attainable ρ, any values above some curve $\lambda = g(\rho)$ being attainable $(g(0) = \frac{1}{2}, g(1) = 1)$. One would further hypothesize that g is an increasing function, but this has not been proved, and its exact form has not been determined. Thompson (1980*b*) investigates the relationship between λ and ρ

under a variety of mating schemes, and thereby delimits the restrictions, but the complete solution remains an open problem.

This problem of correlation in descent has wider implications than simply a determination of the attainable gene-identity-states. The limited paths of descent in a complex genealogy, which give rise to correlations in the descent of genes, are an important factor in restricting the distribution of traits in a small population. Any analysis of the observed distribution of an inherited trait is, implicitly if not explicitly, an analysis of the effect of correlations in descent. We shall therefore return to these correlations in Chapter 4.

2.8   Linked loci and X-linked loci

In this chapter we have considered only independent autosomal loci. The same questions arise for X-linked loci, at which a male individual carries only a single gene. In addition, many genetic characteristics may be controlled by several linked loci, and the questions of joint phenotype distributions for such characteristics are of practical importance.

At an X-linked locus, a male passes his single gene to each female offspring, but does not contribute to male offspring. A female contributes to both male and female offspring, Mendelian segregation operating precisely as for autosomal loci. Thus we may specify the space of elementary gene-identity-states precisely as in Section 2.4, requiring that a male always has two identical genes (a notational convenience for his single one). The method of partition of the total set of states into equivalence classes under transformations T_i is then unaltered and (2.5.3) and (2.5.4) for joint genotype and phenotype distributions continue to hold, the latter with appropriate definitions of penetrance probabilities, possibly different for males and females.

Thus the only essential difference between X-linked and autosomal loci is in the calculation of probabilities of gene-identity-states, since these involve the transmission mechanism, males now receiving *two identical copies* of a maternal gene. The computations are most readily modified by a modification of the genealogical graph. All father–son links may be deleted, and male individuals, other than those in the final set of interest, may be simply removed provided they have no more than one relevant daughter. A single daughter is in effect a *child* of her paternal grandmother. A set of *n* full sisters are as *identical n-plets* with respect to their gene received from this paternal grandmother, the father thus becoming a uniparental twin junction. Computations are then as for autosomal loci.

More theoretical developments are provided by the case of linked loci. Here the space of gene-identity-states must be generalized to specify joint identities at several loci separately, and computation of these gene-identity-state probabilities involves the joint transmission of genes and the recombination parameters for the loci. Haldane (1949) first considered the extension of the IBD concept of Cotterman (1940) and Malecot (1948) to linked loci. Cockerham (1956) discussed the effect of linkage on the covariance between relatives for a quantitative trait, considering in detail the cases of half-sibs and sibs, while a more general formulation was given by Schnell (1963). Both these authors considered only non-inbred relatives. This restriction was removed by Gallais (1974), whose approach combined the single-locus identity states of Gillois (1964) with a *link relationship*, a coefficient expressing the probability that genes at different loci derive from the same ancestral gamete.

Campbell and Elston (1971) instead generalize the single-locus transition-matrix formulation of Li and Sacks (1954) to derive the conditional genotype, and hence phenotype, distribution of a relative. Their restrictions, to non-inbred individuals, allow them to distinguish only seven different identity states for the eight genes of two relatives at two loci. They provide the seven transition matrices giving the conditional genotype of a relative, given that of a *proband* in each identity state. Lange (1974) further extends this analysis of transition probabilites for two linked loci.

Since there are 15 elementary single-locus gene identity states between two individuals, there are in all 225 for two loci (Cotterman, 1940). For non-inbred relatives there are seven identity states at one locus (see Table 2.3(a)), and hence 49 at two. However, as in the single-locus case, not all states are genotypically distinguishable. The problem for two loci is complicated by the fact that, although some identity states are indistinguishable from the point of view of joint phenotype distribution, they may not be so from the point of view of future gene transmission. For example, for some problems it may be necessary to distinguish coupling and repulsion, and in others it is not. Denniston (1975) divides the 49 states into 15 classes, and discusses the relationship between these 15 and the seven of Campbell and Elston (1971).

It is clear that, in theory, gene-identity-states for arbitrary numbers of individuals at arbitrary numbers of loci may be defined. Although, as just described, the definition of *equivalent* identity states may depend on the practical problem in question, for any given definition of equivalence, the distinct gene identity states will again be the equivalence classes under some group of trans- formations on the space of identity states. Thus, conceptually, Section 2.4 may be generalized. However in practice it does not seem to be useful to do so. Even where it would be practicable to examine the structure of the gene-identity-state space, the compu- tation of probabilities of these identity states will normally not be feasible.

3

Gene identity in populations

3.1   **Introduction**

We have seen in the previous chapter how the idea of *identity by descent* provides a useful tool of use in computing certain probabilities concerning pairs (or multiplets) of relatives. Our attention in this chapter will be directed towards the study of complete populations. We will be concerned with the expectations of certain coefficients of identity taken over the population, and of the changes of these values through time. The understanding of such changes is important in the study of evolution, and also in animal breeding, where, for example, one may wish to maintain variability at as high a level as possible.

Our basic tool, then, will be a coefficient of identity averaged over some class of pairs of relatives (e.g. $\psi_S^{(t)}$ will be the coefficient of kinship of a randomly chosen pair of sibs in generation t). Certain features of the population structure can be summarized by a set of such coefficients and relations derived between these coefficients in successive generations. In this way we can extract information about certain types of changes in the population structure. The particular set of coefficients required will be dictated by the model of mating (i.e. whether random, or with certain preferences and/or prescriptions, whether permanent or temporary etc.), and in some cases by the existence of different age-classes.

The advantage of using this approach to the study of population genetic models, *vis-à-vis* a more formal approach through standard Markov chain theory, is twofold. In the first place, one need not specify so completely the features of the model (i.e. a wider class of models can be treated at any stage). Secondly, the entities one is dealing with, i.e. the coefficients of identity, are in some ways more intuitively familar to the geneticist than the latent roots and vectors of Markov chains (although there is an intimate connection with them).

It will be impossible to summarize all of the vast literature on models of populations set up to study the changes of genetic composition. Accordingly we shall discuss a selection which will enable the introduction of a wide variety of important features, and hence the presentation of the techniques involved in using coefficients of identity for population studies.

3.2 Closed haploid populations

In considering haploid populations with only a single locus we have a particular simple example of the use of coefficients of kinship. Suppose we have a population of haploids, with no immigration or emigration (i.e. closed). We characterize the survival and reproduction of the population by two parameters

$$p_S = P \{\text{two randomly selected individuals are sibs}\}$$

and

$$p_{PO} = P\{\text{two randomly selected individuals are parent and offspring}\}$$

these being assumed fixed throughout time. Here, as elsewhere, we take a discrete time model. It is straightforward to see that

$$\psi^{(t)} = p_S\psi_S^{(t)} + p_{PO}\psi_{PO}^{(t)} + (1 - p_S - p_{PO})\psi^{(t-1)}, \qquad (3.2.1)$$

where, as before $\psi_S^{(t)}$ is the coefficient of identity for sibs at time t, $\psi_{PO}^{(t)}$ is that for parent and offspring (the offspring having just been born at time t), and $\psi^{(t)}$ is the population (mean) coefficient of identity. Now, for a haploid population without mutation $\psi_S^{(t)} = \psi_{PO}^{(t)} = 1$; hence, putting $H^{(t)} = 1 - \psi^{(t)}$ we obtain

$$H^{(t)} = (1 - p_S - p_{PO})H^{(t-1)}. \qquad (3.2.2)$$

Accordingly (unless $p_S + p_{PO} = 0$, i.e. there is no reproduction) $H^{(t)} \to 0$ as $t \to \infty$, corresponding to the ultimate fixation of some allele. The rate of loss of variability (or rate of drift) is simply λ, say, where

$$\lambda = (1 - p_S - p_{PO}) \qquad (3.2.3)$$

which corresponds to the dominant latent root of the underlying Markov chain.

It should be noted that (3.2.1) and (3.2.3) are completely general, provided that the complete specification of reproduction and survival is defined by p_S and p_{PO}. This will be the case if all individuals behave in a stochastically identical fashion, but no other restriction is necessary. As an illustration of this latter point we may observe the possibility of introducing interdependence between reproduction and survival, as in Section 3.2 (v).

3.2(i)   *Wright's* (1931) *model*

If the generations do not overlap ($p_{PO} = 0$), and if each offspring has probability $1/N$ of being derived from any particular individual, N being the population size, then we have $p_S = 1/N$, and

$$\lambda = \left(1 - \frac{1}{N}\right). \tag{3.2.4}$$

This model is often used as a reference point for comparing models. Thus, if for some other model we have λ^*, then N_e, the effective population size, is defined by

$$\lambda^* = \left(1 - \frac{1}{N_e}\right)$$

i.e.

$$N_e = \frac{1}{1 - \lambda^*}. \tag{3.2.5}$$

N_e is the size of population which under Wright's model would give the same rate of drift. More general aspects of non-overlapping generation models are treated by Karlin and McGregor (1965).

3.2.(ii)   *Fixed size, non-overlapping generation models*

If $p_{PO} = 0$ so that the generations do not overlap we have $\lambda = 1 - p_S$, and it is useful to reexpress this in terms of the family size distribution. If $f_i = P\{$individual has i offspring$\}$, consider the selection of an individual from some generation. The probability that the individual comes from a family of size i is $if_i/\sum_i if_i$, in which case he has $(i - 1)$ sibs, so that on selecting another individual we have

$$p_S = \frac{\sum_i i(i - 1)f_i}{(N - 1)\sum_i if_i} = \frac{\sigma^2 + \mu^2 - \mu}{\mu(N - 1)}, \tag{3.2.6}$$

where σ^2 = variance of family size, and μ = mean family size. (This presentation follows that of Jacquard (1971).)

Now for a fixed size haploid model, $\mu = 1$, so
$$p_S = \sigma^2/(N-1), \quad \lambda = 1 - \sigma^2/(N-1),$$
and
$$N_e = (N-1)/\sigma^2.$$
We note that for a haploid model $0 \le \sigma^2 \le (N-1)$, the limits being for the cases where each individual produces one offspring, and for that where one, randomly chosen, individual produces all N offspring.

3.2(iii) *Moran's (1958) model*

The earliest model in which the generations overlapped is that due to Moran (1958). In this, at each time point, one individual is chosen to die, and one to produce a single offspring, these selections being made independently. As there is only one new individual produced $p_S = 0$, while

$$p_{PO} = P \{\text{Parent survives}\}$$

$$\times P \{\text{Parent and offspring picked}\,|\,\text{Parent survives}\}$$

$$= \frac{(N-1)}{N} \frac{2}{N(N-1)} = \frac{2}{N^2},$$

i.e. the rate of drift $\lambda = (1 - 2/N^2)$.

Since in this model only one individual is born per unit time it is more relevant to compare λ^N with that of a non-overlapping generation model. λ^N is (for large N) approximately $(1 - 2/N)$, so $N_e \approx N/2$.

3.2(iv) *More general overlapping generation models*

Chia and Watterson (1969) introduced a model in which a random number of individuals die, these being replaced by an equal number of new individuals. A somewhat more general formulation was given by Cannings (1974a), and we follow this latter approach here. Suppose the ith individual in generation $(t-1)$ produces η_i survivors (necessarily $\eta_i = 0$ or 1), and ξ_i offspring. The (η_i, ξ_i) are assumed exchangeable, i.e. individuals behave in a stochastically identical fashion. Then

$$p_S = \frac{E(\xi_i(\xi_i - 1))}{(N-1)} \quad \text{and} \quad p_{PO} = \frac{2E(\xi_i\eta_i)}{(N-1)},$$

(where E denotes expectation) so the rate of drift

$$\lambda = 1 - \frac{E(\xi_i(\xi_i - 1))}{(N-1)} - \frac{2E(\xi_i \eta_i)}{(N-1)}$$

$$= 1 - \frac{E(\theta_i(\theta_i - 1))}{(N-1)},$$

where $\theta_i = \xi_i + \eta_i$ (noting that $E(\eta_i) = E(\eta_i^2)$), so

$$\lambda = E(\theta_i \theta_j) \quad \text{for } i \neq j, \tag{3.2.7}$$

the latter step following by using $E((\sum \theta_i)^2) = N^2$. The use of θ_i, which is the total number of members of the family of the ith individual in the next generation, illustrates the fact that in haploid models, since parent and offspring are identical, we can replace any overlapping generation model by an equivalent non-overlapping one, simply by treating a surviving individual as if it were one of its own offspring (Cannings, 1973).

3.2(v) *Interaction of survival and reproduction*

As an illustration of the flexibility of models encompassed by (iv) above ((i), (ii) and (iii) also being special cases) we introduce the following model. Suppose that one individual is replaced at each point of time. One is picked at random to die, the offspring is to come from this individual with probability q, and from any other particular individual with probability $(1-q)/(N-1)$. Then $p_S = 0$ and $p_{PO} = 2(1-q)/N(N-1)$. Hence $\lambda = 1 - 2(1-q)/N(N-1)$ and $N_e \approx (N-1)/2(1-q)$.

The equation (3.2.1) is valid whether or not the process is time homogeneous, provided that p_S and p_{PO} are interpreted as being the values appropriate at time t. We examine three particular cases of variable sized populations.

3.2(vi) *Cyclic population size*

Suppose, as per Chia (1968), that there is a sequence N_1, $N_2, \ldots, N_k$ of population sizes, the population size then reverting to N_1. Introducing $H_i^{(t)}$ the coefficient of non-identity by descent at time t (when the size is N_i) we have

$$H_i^{(t)} = (1 - p_i)H_{i-1}^{(t-1)}, \text{ from (3.2.2),}$$

where p_i is the appropriate value of $p_S + p_{PO}$. Thus

$$H_{i+k}^{(t)} = \prod_{i=1}^{k} (1-p_i)H_i^{(t-k)}$$

so that the rate of loss of variability per cycle is

$$\lambda = \prod_{i=1}^{k} (1-p_i). \tag{3.2.8}$$

3.2(vii) *Random size* (Karlin 1968a).

Suppose that the size at any time point is n_i with probability r_i, $i = 1, 2, \ldots, k$, there being no correlation between the sizes at different time points. Defining $H_i^{(t)}$ as above we have

$$H_i^{(t)} = \sum_{j=1}^{k} r_j(1-p_{ji})H_j^{(t-1)}, \tag{3.2.9}$$

where p_{ji} equals the value of $p_S + p_{PO}$, appropriate when a population of size n_j gives rise to one of size n_i. Thus

$$\boldsymbol{H}^{(t)} = \boldsymbol{A}\boldsymbol{H}^{(t-1)}, \tag{3.2.10}$$

where

$$\boldsymbol{H}^{(t)} = (H_1^{(t)}, H_2^{(t)}, \ldots, H_k^{(t)})'$$

and

$$\boldsymbol{A} = (a_{ij}), \, a_{ij} = r_j(1-p_{ji}).$$

We should ideally like to find the complete set of latent roots and vectors of $\boldsymbol{A}$. However, we content ourselves here with deriving only the dominant root λ in certain special cases (these and others are discussed by Karlin (1968a)).

Suppose that $p_{ji} = p_j$, as is the case in certain important models. In that case $a_{ij} = r_j(1-p_j)$, and the columns of $\boldsymbol{A}$ are constant. For such a matrix there is a left latent vector given by $\boldsymbol{u} = (u_1, u_2, \ldots, u_k)$, where $u_i = r_i(1-p_i)$, with latent root $\lambda = \sum_i r_i(1-p_i)$. This λ is the dominant latent root from Frobenius theory ($\boldsymbol{A}$ being positive and $\boldsymbol{u}$ being positive guarantee this, see, for example, Karlin and Taylor (1975)). Now since

$$\boldsymbol{u}\boldsymbol{A} = \lambda\boldsymbol{u}$$

and

$$\boldsymbol{H}^{(t)} = \boldsymbol{A}\boldsymbol{H}^{(t-1)},$$

we have

$$\mathbf{u} \cdot \mathbf{H}^{(t)} = \mathbf{u}\mathbf{A}\mathbf{H}^{(-1)}$$

$$= \lambda \mathbf{u} \cdot \mathbf{H}^{(t-1)}$$

and so the function $\mathbf{u} \cdot \mathbf{H}^{(t)}$, which equals $\sum_i r_i(1 - p_i)H_i^{(t)}$, is scaled by a factor λ, the rate of drift, per generation.

In two of the cases considered by Karlin (1968a) $p_{ji} = p_j$, viz, if we have reproduction as for the Wright model, $p_{ji} = 1/N_j$, and if we have a population in which each individual makes a single copy of itself, and the next generation is taken at random from this set of $2N_j$, we have $p_{ji} = 1/(2N_{j-1})$. In fact $p_{ji} = p_j$ if and only if $E(\theta_l\theta_m) = N_i(N_i - 1)g(N_i)$ for some function $g(\cdot)$.

We have therefore $\lambda = \sum_i r_i(1 - p_i) = 1 - \sum_i r_i p_i$, and for a single population size $\lambda = 1 - p$. For Wright's model we obtain, for the special case $r_i = 1/k$, an effective population size given by

$$\frac{1}{N_e} = \left(\sum_i \frac{1}{N_i}\right) \Big/ k$$

and for the second model an effective population size, N_e^*, given by

$$\frac{1}{N_e^*} = \left(\sum_i \frac{1}{2N_i - 1}\right) \Big/ k$$

vis-à-vis $1/(2N - 1)$ for a single population size.

3.2(viii) *Markovian size* (Karlin, 1968a, Chia, 1968)

We have above considered a deterministic sequence of population sizes, and a random one with independence of the sizes in different generations. A reasonable biological situation might require that there were some dependence between successive generations, as provided for in a Markov chain model. Suppose then that there are k possible population sizes, and that the size at some time t is specified by the random variable $N^{(t)}$, with

$$P\{N^{(t+1)} = N_j | N^{(t)} = N_i\} = r_{ij} \qquad i, j = 1, 2, \ldots, k.$$

Then

$$H_i^{(t)} = \frac{\sum_j r_j^{(t-1)} r_{ji} H_j^{(t-1)}(1 - p_{ji})}{\sum_j r_j^{(t-1)} r_{ji}}, \qquad (3.2.12)$$

where $r_j^{(t)} = P\{N^{(t)} = N_j\}$, $r_j^{(t-1)} r_{ji} / \sum r_j^{(t-1)} r_{ji}$ being the probability that $N^{(t-1)} = N_j$ given $N^{(t)} = N_i$. As our interest is in the behaviour of $H_i^{(t)}$'s we take the r_js at their equilibrium values, $\hat{r}_j$s say. Since $\hat{r}_i = \sum_l \hat{r}_l r_{li}$ we have, from (3.2.12),

$$\hat{r}_i H_i^{(t)} = \sum_j r_{ji}(1 - p_{ji})\hat{r}_j H_j^{(t-1)} \tag{3.2.13}$$

and the latent roots of relevance are those of the matrix A, where $a_{ij} = r_{ji}(1 - p_{ji})$.

3.3 Geographically subdivided haploid populations

We have only considered single populations in the discussion above. However, it is important to study the behaviour in cases where a number of subpopulations are connected by immigration and emigration.

We suppose there are Q colonies, and that in each generation K individuals emigrate from the ith colony to the jth colony, for each pair of i and j. If each colony is of size N, then the total population has QN individuals, and $R = N - (Q-1)K$ is the number of non-migrants per colony. This model is due to Moran (1959), except that we shall not adopt his assumption of Wright-type reproduction, but leave this unspecified (as per Cannings (1975)), except that it is assumed that each subpopulation reproduces similarly (in a stochastic sense), and independently.

Now a complete specification of the coefficients of kinship require that we define $H_{ij}^{(t)}$, $i, j = 1, 2, \ldots, Q$, the coefficient of non-identity for individuals drawn at random one from the ith and one from the jth subpopulations. However, the assumptions of the model imply that after one generation there are only two distinct Hs, viz., $H_W^{(t)}$ and $H_B^{(t)}$ say, where $H_W^{(t)} = H_{ii}^{(t)}$, $i = 1, 2, \ldots, Q$ and $H_B^{(t)} = H_{ij}^{(t)}$, $i \neq j, i, j = 1, 2, \ldots, Q$, W and B indicating *within* and *between*. Our investigation can concentrate on these two values alone. We should note that this reduction to two coefficients occurs because we are dealing with expectations; there will in practice be differences between populations.

The relationships for the Hs are given by

$$\begin{aligned} H_W^{(t)} &= u_W(1 - p_S)H_W^{(t-1)} + (1 - u_W)H_B^{(t-1)} \\ H_B^{(t)} &= v_W H_W^{(t-1)} + (1 - v_W)H_B^{(t-1)}, \end{aligned} \tag{3.3.1}$$

where

$u_W = P$ {two individuals were in the same subpopulation prior to migration|they are in same subpopulation after migration}

$$= \frac{R(R-1)+(Q-1)K(K-1)}{N(N-1)},$$

and v_W is the similar probability conditional on the individuals being in different populations after migration, thus

$$v_W = \frac{2RK+(Q-2)K^2}{N^2}.$$

Here p_S is the probability that two individuals taken from the same subpopulation before migration are sibs.

The latent roots are therefore given by the roots of

$$\lambda^2 - \lambda\{u_W(1-p_S)+(1-v_W)\} \\ +\{u_W(1-v_W)(1-p_S)-v_W(1-u_W)\} = 0. \quad (3.3.2))$$

If $v_W = 0$, i.e. if $K = 0$, $R = N$ so $u_W = 1$ and (3.3.2) reduces to

$$\lambda^2 - \lambda(2-p_S)+1-p_S = 0,$$

so the largest latent root is 1, and the rate of drift is given by the other root $\lambda = (1-p_S)$. This case corresponds to a number of subpopulations with no intermigration, and the rate of drift, $(1-p_S)$, will of course be affected by the subdivision as p_S will tend to be larger in smaller subpopulations.

More complex patterns of migration can be accommodated by introducing further coefficients of non-identity. For example a set of subpopulations arranged in a circle, with migration dependent only on the distances between these subpopulations, could be dealt with by introducing $H_0^{(t)}, H_1^{(t)}, \ldots, H_k^{(t)}$, where $H_i^{(t)}$ is the coefficient of non-identity of two individuals drawn from populations i apart.

3.4   Open haploid populations

An additional complexity is added when we consider an open population, i.e. one with immigration from some outside source, or with mutation. In these cases, as we shall see below, the population will not go to fixation of an allele, but settle down to some equilibrium distribution of allele frequency. We shall limit our consideration of this class of models to certain special cases, which illustrate the type of approach needed.

3.4.(i) *Wright's island model* (Wright, 1943)

Suppose we are interested in the population of an island, and that there is some immigration from the mainland. We ignore emigration to the mainland, which might be absent owing to prevailing wind or tide condition, or be considered irrelevant if the mainland population is very large compared with the island's. Suppose that the proportion of the island's population derived from the mainland is γ in each generation, and denote by $H_{AB}^{(t)}$ the coefficient of non-identity for individuals taken from locations A and B, where A, B = I, for island, or M, for mainland. Then

$$\left.\begin{aligned} H_{II}^{(t)} &= (1-\gamma)^2(1-p_S)H_{II}^{(t-1)} + 2\gamma(1-\gamma)H_{IM}^{(t-1)} + \gamma^2 H_{MM}^{(t-1)} \\ H_{IM}^{(t)} &= (1-\gamma)H_{IM}^{(t-1)} + \gamma H_{MM}^{(t-1)} \\ H_{MM}^{(t)} &= H_{MM}^{(t-1)}. \end{aligned}\right\} \quad (3.4.1)$$

The final expression in (3.4.1) corresponds to the assumption regarding the mainland population. Now the matrix implicit in (3.4.1) is upper triangular, and hence the three latent roots are simply the diagonal elements, 1, $(1-\gamma)$ and $(1-\gamma)^2(1-p_S)$. These latent roots can be identified as the rates of change of $H_{MM}^{(t)}$, $H_{IM}^{(t)}$ and $H_{II}^{(t)}$ respectively, and so the rate of relevance here is $\lambda = (1-\gamma)^2(1-p_S)$. The asymptotic values of $H_{MM}^{(t)}$ and $H_{MI}^{(t)}$ are simply $H_{MM}^{(0)}$, while that of $H_{II}^{(t)}$, $H_{II}^{(\infty)}$ say, is given by

$$H_{II}^{(\infty)} = \lambda H_{II}^{(\infty)} + \{2\gamma(1-\gamma) + \gamma^2\}H_{MM}^{(0)}$$

so

$$H_{II}^{(\infty)} = \frac{\{2\gamma(1-\gamma) + \gamma^2\}H_{MM}^{(0)}}{1 - (1-\gamma)^2(1-p_S)}. \quad (3.4.2)$$

In Wright's formulation attention was focused on the variance of gene-frequency on the island, while the migration from the mainland was assumed to be with a constant gene-frequency, $\bar{q}$ say. If we regard $\bar{q}$ as the frequency of individuals all descended from one common ancestor, and $(1-\bar{q})$ as the frequency of those descended from some other, then $H_{MM}^{(0)} = \bar{q}(1-\bar{q})$. Writing $V_I^{(\infty)}$ for the asymptotic variance of gene-frequency on the island, and r^∞ for the gene-frequency, we have

$$H_{II}^{(\infty)} = E(r^\infty(1-r^\infty))$$
$$= -V_I^{(\infty)} + \bar{q}(1-\bar{q}),$$

and obtain

$$V_1^{(\infty)} = \frac{\bar{q}(1-\bar{q})p_S}{1-(1-p_S)(1-\gamma)^2}.$$ (3.4.3)

This equation is the usual form given as the measure of the structure of the island population.

The study of the effects of various migration patterns has been pursued by a number of authors. In particular Kimura and Weiss (1964) and Weiss and Kimura (1965) have investigated a linear chain of islands, Bodmer and Cavalli-Sforza (1968) have set up a general migration matrix model, while Maruyama (1969, 1970a, b, c) has studied a variety of interesting cases.

3.4(ii) *Mutation models*

In the island model the input of new (to the island) genes occurs with a given gene-frequency, and does not depend on the present frequency on the island. In the mutation models this is not the case.

Suppose that there are two alleles A_1 and A_2 and that during reproduction an A_1 parent produces an A_2 progeny with probability γ, and an A_2 produces an A_1 with rate β. Now, if our interest is in allele frequency then we need to deal with coefficients of non-identity, by type, rather than by descent (see Section 4.1). In addition we need to introduce $X^{(t)}$, the number of A_1 genes at time t. Then, taking the case of non-overlapping generations, and independence of the mutation of different individual progeny, we have

$$H^{(t)} = E\left\{ p_S\left[\frac{X^{(t-1)}}{N} 2\gamma(1-\gamma) + \frac{(N-X^{(t-1)})}{N} 2\beta(1-\beta) \right] \right.$$

$$+ (1-p_S)\left[\frac{X^{(t-1)}(X^{(t-1)}-1)}{N(N-1)} 2\gamma(1-\gamma) \right.$$

$$+ \frac{2X^{(t-1)}(N-X^{(t-1)})}{N(N-1)} (\gamma\beta + (1-\gamma)(1-\beta))$$

$$\left. \left. + \frac{(N-X^{(t-1)})(N-X^{(t-1)}-1)}{N(N-1)} 2\beta(1-\beta) \right] \right\}$$

$$= E\{ p_S f_1^{(t-1)} + (1-p_S)f_2^{(t-1)} \} \text{ say,}$$

where E denotes expectation with respect to the $X^{(t-1)}$ random variable. Rearranging, and using

$$H^{(t-1)} = \frac{2X^{(t-1)}(N - X^{(t-1)})}{N(N-1)}$$

we obtain

$$H^{(t)} = H^{(t-1)}\{(1 - p_S)(1 - \gamma - \beta)^2\}$$
$$+ 2\gamma(1-\gamma)E\left\{\frac{X^{(t-1)}}{N}\right\} + 2\beta(1-\beta)E\left\{\frac{N - X^{(t-1)}}{N}\right\}.$$

(3.4.5)

The incorporation of overlapping generations into the model gives

$$H^{(t)} = E\left\{p_S f_1^{(t-1)} + p_{PO}\left[\frac{X^{(t-1)}}{N}\gamma + \frac{(N - X^{(t-1)})}{N}\beta\right]\right.$$
$$\left. + (1 - p_S - p_{PO})f_2^{(t-1)}\right\}$$

(3.4.6)

which reduces to

$$H^{(t)} = H^{(t-1)}\{(1 - p_S - p_{PO})(1 - \gamma - \beta)^2\}$$
$$+ \{(1 - p_{PO})2\gamma(1 - \gamma) + p_{PO}\gamma\}E\left\{\frac{X^{(t-1)}}{N}\right\}$$
$$+ \{(1 - p_{PO})2\beta(1 - \beta) + p_{PO}\beta\}E\left\{\frac{(N - X^{(t-1)})}{N}\right\}.$$

(3.4.7)

Accordingly the rates of drift are $(1 - p_S)(1 - \gamma - \beta)^2$ and $(1 - p_S - p_{PO})(1 - \gamma - \beta)^2$, respectively, or simply the latter with $p_{PO} = 0$ for the former model.

Now in both models we have

$$E\left(\frac{X^{(t)}}{N}\right) \to \frac{\beta}{\gamma + \beta} \quad \text{as } t \to \infty,$$

and hence

$$H^{(t)} \to \frac{(1 - p_{PO})\{2\gamma\beta(2 - \gamma - \beta)\} + p_{PO}2\gamma\beta}{(\gamma + \beta)\{1 - (1 - p_S - p_{PO})(1 - \gamma - \beta)^2\}} \quad \text{as } t \to \infty.$$

A somewhat different model results if one considered alleles as specified by their nucleotide sequence. In this case the number of alleles is so vast that each mutation may be assumed to produce a completely new allele, or at least one not presently in the population. Supposing that the rate of mutation is δ,

$$
\begin{aligned}
H^{(t)} &= p_S\{1-(1-\delta)^2\}+p_{PO}\delta \\
&\quad +(1-p_S-p_{PO})[H^{(t-1)}+(1-H^{(t-1)})\{1-(1-\delta)^2)\}] \\
&= H^{(t-1)}(1-p_S-p_{PO})(1-\delta)^2 \\
&\quad +\{1-(1-\delta)^2\}+p_{PO}[\delta-\{1-(1-\delta)^2\}]. \quad\quad (3.4.8)
\end{aligned}
$$

The rate drift is now $(1-p_S-p_{PO})(1-\delta)^2$ and

$$
H^{(t)} \to \frac{\{1-(1-\delta)^2\}+p_{PO}[\delta-\{1-(1-\delta)^2\}]}{1-(1-p_S-p_{PO})(1-\delta)^2} \quad \text{as } t\to\infty.
$$

For the special case where $p_{PO}=0$, and $p_S=1/N$ (i.e. for Wright's model), the two allele model gives

$$
H^{(\infty)} \approx \frac{4\gamma\beta N}{(\gamma+\beta)\{1+2N(\gamma+\beta)\}}
$$

and the infinite allele model gives

$$
H^{(\infty)} \approx \frac{2N\delta}{1+2N\delta},
$$

where it is assumed that N is large and γ, β, δ small (i.e. we neglect γ/N, β/N, δ/N, γ^2 etc.). We note that if $\gamma=\beta=\delta/2$ then for the latter model $H^{(\infty)}$ is exactly double that of the former (subject of course to the approximations made).

3.5   Higher order coefficients for haploid populations

Instead of limiting ourselves to pairs of individuals we may wish to extend our treatment to multiplets, as is discussed in the previous chapter. Here we will discuss triplets so as to make the ideas familiar, and then discuss an application of these coefficients.

Consider then a triplet of individuals in a haploid population. In an overlapping generation model there are nine distinct specifications of the relevant relationships between the three individuals, in terms of the individuals of the previous generation. However, it is only important here to distinguish between triplets which are not in any way related as parent and offspring (i.e. three survivors; two survivors and the progeny of some other individual; one survivor, and two progeny of other individuals who are not themselves sibs; or three progeny who are not sibs.); triplets which involve a parent–offspring pair or a sib pair, and an unrelated individual; and triplets who are all in the same family (i.e. three sibs, or an individual and two of its progeny). We denote the probability of these three cases by p_3, p_2 and p_1. Similarly we define coefficients of kinship.

$\psi_I^{(t)} = P$ {three randomly selected individuals in generation t are identical by descent}, $\psi_{II}^{(t)} = P$ {three randomly selected individuals in generation t have two identical by descent, and one non-identical by descent}, and $\psi^{(t)} = $ the coefficient of identity for two randomly selected individuals. Then

$$\left.\begin{array}{l} \psi_I^{(t)} = p_1 + p_2\psi^{(t-1)} + p_3\psi_I^{(t-1)} \\ \psi_{II}^{(t)} = p_2(1-\psi^{(t-1)}) + p_3\psi_{II}^{(t-1)} \\ \psi^{(t)} = (p_{PO}+p_S) + (1-p_{PO}-p_S)\psi^{(t-1)}. \end{array}\right\} \quad (3.5.1)$$

Adding the first two, and putting $H_3^{(t)} = 1 - \psi_I^{(t)} - \psi_{II}^{(t)}$, we obtain

$$H_3^{(t)} = p_3 H_3^{(t-1)},$$

$H^{(t)}$ being the probability that three individuals are mutually non-identical by descent.

More generally (following Felsenstein, 1971a) we can see that with k individuals, with $p_k = P\{k$ individuals are non-related when generation t and $(t-1)$ are taken into account}, and $H_k^{(t)}$ the probability of mutual non-identity of k randomly selected individuals, that

$$H_k^{(t)} = p_k H_k^{(t-1)}.$$

Moreover (Felsenstein, 1971a) has shown that

$$H_{ij}^{(t)} = \sum_{k=1}^{N} \Gamma_{ik} H_{kj}^{(t-1)}, \quad (3.5.2)$$

where $H_{ij} = P\{i$ individuals are of j different types}, and $\Gamma_{ik} = P\{i$ individuals have k parents}. Above $\Gamma_{33} = p_3$, $\Gamma_{32} = p_2$ and $\Gamma_{31} = p_1$. Similar equations have been given by Kempthorne (1968). Felsenstein (1971a) and Karlin (1968b, c) use the coefficients to find the rate of elimination of alleles from a multi-allelic population (see also Section 4.3). It can be shown that the rate of elimination of an allele when there are presently k alleles is given by $E(\Pi_{i=1}^{k}\theta_i)$, where the θ_is are as defined in Section 3.2(iv).

3.6 Diploid models: introduction

We now turn to models in which the individuals are diploids, and we consider in the main a single autosomal locus. The consideration of diploids, *vis-à-vis* haploids, requires that we take account of a variety of additional factors, e.g. selfing, marriage rules etc. By a marriage rule we simply mean a rule specifying the structure of pairing of male and female within the population. These rules may be simple, e.g. that individuals mate entirely at random, or that they practice monogamy having obtained a spouse at random, or they may be much more complex, e.g. requiring that individuals with a certain degree of biological, or affinal, relatedness do, or do not, marry (see Chapter 1). In human populations other factors such as age, geographic location, and social position would play a part, but will be omitted here. Also excluded are models in which marriage rules are dependent on the genetic type of the potential spouses.

3.7 A basic diploid model

Our basic model is one in which no account is taken of relatedness in determining matings, so that individuals are considered indistinguishable, except for their sex, for purposes of pairing. Thus the parents of any individual will be randomly selected if he has two parents, a randomly selected individual if he has only one.

In deriving expressions for the relevant coefficients of non-identity at time t we need to identify all the possible types of pairs of individuals. The factor which determines a particular type is the relationship between the two individuals, where relationship is limited to that created by the step from $(t-1)$ to t, i.e. not differentiating between individuals at time $(t-1)$. In our basic model, in which selfing is permitted, there are seven different types of pairs: a pair of sibs each having been produced by selfing, with frequency p_{S^*} say; a pair of sibs each produced by a mating between the same pair of parents, with frequency p_S; a pair of half-sibs in which one was produced by selfing, the other by the mating of the other's single parent and some other, with p_{HS^*}; a pair of half-sibs each produced by a mating, and with one parent in common, with frequency p_{HS}; a parent and an offspring, produced from that parent by selfing, with frequency p_{PO^*}; a parent and an offspring, produced from that parent's mating with another adult, with frequency p_{PO}; two unrelated individuals (two offspring with no parents in common, a survivor and an offspring not produced by the survivor, or two survivors), with frequency $(1 - p_{S^*} - p_S - p_{HS^*} - p_{HS} - p_{PO^*} - p_{PO})$. As before all of these six parameters are fixed through time.

We thus have

$$H^{(t)} = \sum_R p_R H_R^{(t)} + (1 - \sum_R p_R) H^{(t-1)}, \qquad (3.7.1)$$

where R stands for S^*, S, HS^*, HS, PO^* and PO. Using the results of Section 2.2 on the expressions for the various $H_R^{(t)}$, we obtain

$$\begin{aligned} H^{(t)} &= (p_S + p_{HS^*} + p_{PO})\{\tfrac{1}{4}(2H^{(t-1)} + H_I^{(t-1)})\} \\ &\quad + (p_{PO^*} + p_{S^*})\{\tfrac{1}{2}H_I^{(t-1)}\} + p_{HS}\{\tfrac{1}{8}(6H^{(t-1)} + H_I^{(t-1)})\} \\ &\quad + (1 - p_S - p_{HS^*} - p_{PO} - p_{PO^*} - p_{S^*} - p_{HS})H^{(t-1)} \\ &= (1 - \tfrac{1}{2}T)H^{(t-1)} + \tfrac{1}{4}T H_I^{(t-1)}, \qquad (3.7.2) \end{aligned}$$

where $T = (p_S + p_{HS^*} + p_{PO}) + 2(p_{PO^*} + p_{S^*}) + \tfrac{1}{2}p_{HS}$. The term T has a fairly simple interpretation. Consider, for example the HS* relationship; the parents of one individual are A and B say, and of the other A alone, thus the probability that a randomly selected parent of one is identical (i.e. the same individual) with a randomly selected parent of the other is $\tfrac{1}{2}$. Similarly we obtain for the relationship S the value $\tfrac{1}{2}$, for S* the value 1, and for HS the value $\tfrac{1}{4}$. For the relationships PO and PO* we simply take the parent *vis-à-vis* the parent or parents of the offspring, giving values $\tfrac{1}{2}$ and 1 respectively. Thus T is a measure of the identity of individuals in the previous generation.

In addition to (3.7.2.) we need another recurrence expressing $H_I^{(t)}$ in terms of coefficients at $(t-1)$. $H_I^{(t)}$ is the coefficient of inbreeding of an individual. In determining this value we need only differentiate three cases: the individual is a survivor from time $(t-1)$, with frequency p_A; the individual is an offspring produced by selfing, with frequency p_B; the individual is an offspring produced by a mating of two individuals, with frequency $p_C = 1 - p_A - p_B$. Hence we obtain

$$\begin{aligned} H_I^{(t)} &= p_A H_I^{(t-1)} + \tfrac{1}{2}p_B H_I^{(t-1)} + (1 - p_A - p_B)H^{(t-1)} \\ &= (p_A + \tfrac{1}{2}p_B)H_I^{(t-1)} + (1 - p_A - p_B)H^{(t-1)}. \qquad (3.7.3) \end{aligned}$$

The equations (3.7.2) and (3.7.3) enable us to determine the appropriate rates of change of the coefficients of non-identity.

In particular when the generations are non-overlapping and no selfing occurs (3.7.3) reduces to

$$H_I^{(t)} = H^{(t-1)} \tag{3.7.4}$$

so that the latent roots are given by the roots of

$$\lambda^2 - \lambda(1 - \tfrac{1}{2}T) + \tfrac{1}{4}T = 0. \tag{3.7.5}$$

In the case where T is small, $\lambda \approx 1 - \tfrac{1}{4}T$. More generally, if λ_1 and λ_2 are the latent roots, corresponding to (3.7.2) and (3.7.3), then

$$(H^{(t)} - \lambda_i H^{(t-1)}) = \lambda_j (H^{(t-1)} - \lambda_i H^{(t-2)})$$

with $(i, j) = (1, 2)$ or $(2, 1)$, we have

$$H^{(t)} - \lambda_i H^{(t-1)} = \lambda_j^t (H^{(0)} - \lambda_i H^{(-1)}), \tag{3.7.6}$$

this pair of expressions allowing $H^{(t)}$ to be expressed in terms of $H^{(0)}$ and $H^{(-1)}$.

This basic model includes various important special cases (only non-zero p_R values being specified, i.e. all others are zero).

3.7(i) *Repeated sib-mating* (see also Section 4.2(i))

If there is just one male and one female in the population in each generation, and generations are non-overlapping, then $p_S = 1$, so $T = 1$, and

$$H^{(t)} = \tfrac{1}{2}H^{(t-1)} + \tfrac{1}{4}H^{(t-2)},$$

and the latent roots are $\tfrac{1}{4}(1 \pm \sqrt{5})$.

3.7(ii) *Wright's* (1931) *dioecious model*

Suppose there are N_1 males and N_2 females, non-overlapping generations, and each offspring has as parents a randomly selected male and female independently of all other offspring. In this case, $p_S = 1/N_1 N_2$ and $p_{HS} = (N_1 + N_2 - 2)/N_1 N_2$, so $T = (N_1 + N_2)/2N_1 N_2$, and the rate of drift $\lambda \approx 1 - \tfrac{1}{4}T = 1 - \{(N_1 + N_2)/8N_1 N_2\}$, which if $N_1 = N_2 = \tfrac{1}{2}N$ gives $\lambda = 1 - (1/2N)$. Thus a population of N, with equal number of males and females, has the same rate of drift as a haploid population of $2N$. Thus it is only the number of haploids which is important in Wright's model. Thus $N_e = 2N$ for this model if $N_1 = N_2 = \tfrac{1}{2}N$, and $N_e = 8N_1 N_2/(N_1 + N_2)$ otherwise.

3.7(iii) *Random polygamy*

A more general model of Cannings (1974*b*) allows for polygamy, in one or both sexes, and for a fairly general offspring distribution. Supposing that males produce a number of offspring with mean N/N_1 and variance σ_1^2, and females produce with mean N/N_2 and variance σ_2^2, then (with non-overlapping generations, and no selfing) (from (3.2.6))

$$T = p_S + \tfrac{1}{2}p_{HS} = \tfrac{1}{2}\{P\{\text{mother in common}\} + P\{\text{father in common}\}\}$$

$$= \frac{(N_1\sigma_1^2 + N_2\sigma_2^2 - 2N + N^3/N_1N_2)}{2N(N-1)}, \qquad (3.7.7)$$

where, as before, $N_1 =$ number of males and $N_2 =$ number of females. This result was originally given by Moran and Watterson (1959) for a somewhat more restricted model. For Wright's model above, $\sigma_1^2 = N(N_1 - 1)/N_1^2$ and $\sigma_2^2 = N(N_2 - 1)/N_2^2$, so $T = N/2N_1N_2$ as given in Section 3.7(ii). Some more complex cases are discussed in Cannings (1974*b*), including models where the distribution of the number of mates needs to be specified. For example, suppose one has a system of polygamy, with each female being assigned at random to a male independently of all other females, and (assuming $N_1 = N_2$), each female producing offspring with mean number 2 and variance σ^2. We then have $T = (\sigma^2 + 1)/(N-1)$, which contrasts with $T = (\sigma^2 + 2)/2(N-1)$ obtained for a completely monogamous system. The effect of polygamy can thus be quite marked. For these models variability persists longer in the monogamous system.

The above special cases have been non-overlapping models with no selfing. We now allow selfing, but retain non-overlapping generations.

3.7(iv) *Wright's (1931) monoecious model*

Suppose that there are N individuals, and the parents of any offspring are a randomly chosen pair, where sampling is with replacement. Then $p_B = 1/N$, $p_C = (N-1)/N$, $p_S = 2(N-1)/N^3$, $p_{S^*} = 1/N^3$, $p_{HS^*} = 4(N-1)/N^3$, and $p_{HS} = 4(N-1)(N-2)/N^3$, and $T = 2/N$ (as can be seen directly from the definition of T). In this case (3.7.2) and (3.7.3) reduce to

$$H^{(t)} = H_I^{(t-1)} \frac{1}{2N} + H^{(t-1)}\left(1 - \frac{1}{N}\right)$$

and

$$H_I^{(t)} = H_I^{(t-1)} \frac{1}{2N} + H^{(t-1)}\left(1 - \frac{1}{N}\right),$$

so that $H^{(t)} = H_I^{(t)}$. We then have $\lambda = 1 - (1/2N)$, and this special case and 3.7 (ii) give the same rate of drift.

Our final examples have overlapping generations, the first with selfing, the second without.

3.7(v) *Moran's model 1*

As a generalization of his overlapping generation model for haploids, Moran (1962) introduced two diploid models. In each a randomly selected individual dies, and is replaced by a single offspring. In the model with selfing the offspring is produced from a randomly chosen pair, selected with replacement (i.e. allowing selfing). This model is thus an overlapping generation analogue of case 3.7(iv) above. We have $p_{PO} = 4(N-1)/N^3$ and $p_{PO*} = 2/N^3$, so $T = 4/N^2$, while $p_A = (N-1)/N$, $p_B = 1/N^2$ and $p_C = (N-1)/N^2$. Hence (3.7.2) and (3.7.3) become

$$H^{(t)} = \left(1 - \frac{2}{N^2}\right) H^{(t-1)} + \frac{1}{N^2} H_I^{(t-1)} \tag{3.7.8}$$

and

$$H_I^{(t)} = \frac{(N-1)}{N^2} H^{(t-1)} + \left(1 - \frac{1}{N} + \frac{1}{2N^2}\right) H_I^{(t-1)}, \tag{3.7.9}$$

which, although of different form to the recurrences given by Moran (1958) (in terms of genotype frequencies), yields the same quadratic equation for λ. The appropriate rate is approximately $\{1 - (1/N^2\}$ per birth, or $\{1 - (1/N)\}$ per generation, as compared with $\{1 - (1/2N)\}$ for Wright's model.

3.7(vi) *Moran's model II*

In this second diploid model Moran (1962) has N_1 males and N_2 females. At each stage an individual was selected at random to die, and was replaced by a random mating of a male and female, the offspring having the same sex as the individual which had died.

This model is not encompassed within our present set of equations. In this model there is an interaction between the death event and the birth event, namely the requirement exists that the offspring should be of the same sex as the individual who died. This interaction means that, for example, the probability that two individuals are parent and offspring will depend on the sexes of those individuals. Thus (3.7.2) needs to be replaced by expressions which are different for each possible pair of sexes for two randomly selected individuals. In addition to these three expressions, we will need to replace (3.7.3) by two expressions, one for males and one for females.

We shall not develop these equations here, as the number of parameters involved is relatively large, and an understanding of the ideas involved can be given within the more limited scope of (3.7.2) and (3.7.3).

3.8 Marriage rules based on relationship

In human societies (see for example Chapter 1) marriages are not made at random with respect to age, degree of relationship, membership of clan, and various other factors, and similar effects are found in some animal populations. We now present a model which allows the incorporation of one such marriage rule, in a non-overlapping generation situation.

We suppose that the only degree of relationship which has any effect on marriage chances for two individuals is that of brother–sister. We describe the population behaviour by three parameters, p_S, q and r, where

$$p_S = P \text{ \{two randomly selected individuals are sibs\}},$$
$$q = P \text{ \{a randomly selected male and female}$$
$$\text{are sibs|they are not spouses\}},$$

and

$$r = P \text{ \{a randomly selected male and female}$$
$$\text{are sibs|they are spouses\}}.$$

We use the following coefficients of non-identity: H, H_I, H_S, H_{NS}, H_{SP}, $H_{\overline{SP}}$, being for randomly selected individuals, for a single individual, for sibs, for non-sibs, for spouses, for a male and female who are not spouses. Working with these we obtain

$$\left.\begin{array}{l} H^{(t)} = p_S H_S^{(t)} + (1 - p_S)H_{NS}^{(t)} \\ H_S^{(t)} = \frac{1}{4}(2H_{SP}^{(t-1)} + H_I^{(t-1)}) \\ H_{NS}^{(t)} = \frac{1}{2}(H^{(t-1)} + H_{SP}^{(t-1)}) \\ H_{SP}^{(t)} = rH_S^{(t)} + (1 - r)H_{NS}^{(t)} \\ H_{\overline{SP}}^{(t)} = qH_S^{(t)} + (1 - q)H_{NS}^{(t)} \\ H_I^{(t)} = H_{SP}^{(t-1)}. \end{array}\right\} \tag{3.8.1}$$

Working with these we obtain

$$4H^{(t+1)} - \{4(1-p) + 2r\}H^{(t)} - (2p - r)H^{(t-1)} - (p - r)H^{(t-2)} = 0, \tag{3.8.2}$$

where $p = (p_S + q)/2$.

We examine three special cases.

3.8(i)   *No preferences*

The effect of a sib-mating preference, or avoidance, is removed by taking $q = r = p_S = p$. Hence (3.8.2) becomes

$$4H^{(t+1)} - (4 - 2p)H^{(t)} - pH^{(t-1)} = 0$$

so $4\lambda^2 - 2(2 - p)\lambda - p = 0$ gives the latent roots of relevance here. This equation is equivalent to (3.7.2) with $p = T$.

3.8(ii)   *Small parameters*

If p_S, q and r, and hence p, are small, which will occur for large populations, we can obtain the dominant latent root using Newton's method. We obtain $\lambda \approx 1 - \frac{1}{4}p$, as given for the case with no preferences, or avoidance, though here p has a different interpretation from previously.

3.8(iii)   *Prohibition of sib-mating*

Jacquard (1971) considered the case where $r = 0$ when (3.8.2) becomes

$$4H^{(t+1)} - 4(1-p)H^{(t)} - 2pH^{(t-1)} - pH^{(t-1)} = 0. \tag{3.8.3}$$

For p small we get $1 - \frac{1}{4}p$ again.

Another case of interest occurs when $p_S = 0$, and $q = 1$. In this case we have a regular inbreeding system in which two pairs reproduce, the next generation consisting of two pairs, each formed by taking an offspring from each of the pairs of the previous generation. This system is called the double-first-cousin mating system (Section 4.2(iii)), and $p = \frac{1}{2}$ so we have $8\lambda^3 - 4\lambda^2 - 2\lambda - 1 = 0$, and the largest latent root ≈ 0.9196. This system is illustrated in Figure 3.8(a).

3.9 Marriage rules based on clan system

An alternate approach to the study of marriage rules, and one which corresponds to the practice of many primitive groups, is to base marriage probabilities on clan membership. The set of individuals available for marriage is defined not in terms of biological relationship, but in terms of membership of some particular group. For example, suppose every member of a tribe belongs

Figure 3.8(a). Double-first-cousin mating system.

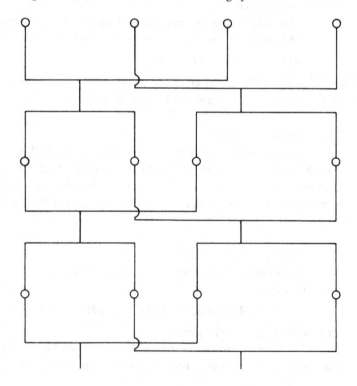

either to group A or to group B, a member of group A must find a spouse from group B, and vice-versa, and the children of a marriage are assigned to their father's group. This system effectively excludes the marriage of siblings, although it is not based on biological relationship directly. A variety of such systems exist, and have been studied by anthropologists. (See, for example, Fox (1967).)

Cannings and Skolnick (1975) treated a variety of such systems, but we shall deal here in detail only with one particularly simple type, namely the two-group exogamous system, as described above. We introduce two coefficients of non-identity, H_W and H_B being the coefficients of non-identity for two individuals taken from within the same group, and taking one from each group, respectively; H_I being defined as before. We denote by p_S the probability that two individuals from the same group are sibs. Then

$$H_W^{(t)} = \tfrac{1}{4}p_S(2H_B^{(t-1)} + H_I^{(t-1)}) + \tfrac{1}{2}(1 - p_S)(H_B^{(t-1)} + H_W^{(t-1)})$$
$$H_B^{(t)} = \tfrac{1}{2}(H_W^{(t-1)} + H_B^{(t-1)}) \qquad (3.9.1)$$

and

$$H_I^{(t)} = H_B^{(t-1)}.$$

From there we obtain

$$8H_W^{(t+2)} - 4(2 - p_S)H_W^{(t+1)} - 2p_S H_W^{(t)} - p_S H_W^{(t-1)} = 0.$$

This is exactly the same expression as (3.8.3) if we put $p_S = 2p$, so that the structure of this model, and the previous one based on relationship, as revealed by the coefficients of non-identity, is the same. Thus the exclusion of sib-mating, whether by direct prohibition, or indirectly through clan rules, has the same effect for genetic drift.

3.10   Conclusion

We have tried to show in this chapter how the study of the changes of gene-frequency distributions is facilitated by the use of coefficients of non-identity. We have of course omitted many models of interest, and many factors of importance. For example, we have omitted any considerations of models in which there is aging. Thus all our models treat the individuals as being indistinguishable demographically; they have, potentially, the same birth and death rate irrespective of how old they are. This enables us to avoid the complexities inherent in keeping track of the age of each individual, but in fact the method is capable of handling more demographically reasonable models. Felsenstein (1971b) has adopted the above technique, introducing a number of age-classes, and taking $H_{ij}^{(t)}$ as the coefficient of non-identity for the ith and jth age-classes at time t. Proceeding, as we have in this chapter, to find recurrence relationships for the $H_{ij}^{(t)}$, Felsenstein finds an approximate expression for the rate of drift. The addition of the age-classes, and the corresponding age-specific birth and death rates, adds to the complexity of the equations obtained, but the basic methodology is the same as set out above.

4

Analysis of population variability

4.1 Gene identity and allelic variability

In the two previous chapters we have defined coefficients of gene identity by descent and considered their change under a variety of population models. In this chapter we consider the evolution of allelic and genotypic variability observable within and between populations. We shall consider only models without mutation, for our interest is in structure and short-term evolution, rather than speciation and long-term selection where mutation also plays an important rôle. There is a close parallel between gene identity and allelic likeness. In the absence of mutation, IBD genes (Section 2.1) are necessarily of like type, and, in the further absence of selection, processes of gene non-identity and allelic difference parallel each other *in expectation*, Much of the methodology of Chapter 3 for analysis of non-identity is applicable equally to non-likeness of allelic type (see Section 3.4(ii), for example).

However, Wright (1965) has cautioned against a too facile identification of the concepts. Although correlation of type between uniting gametes relative to a set of founders is equal to the probability of identical by descent from those same founders, this is not usually the relevant reference frame; allelic variability is often assessed relative to a current pool of populations. Allelic correlations may also, like gene-identity, be defined relative to a specified pedigree structure. Such correlations are necessarily nonnegative, since descent of a particular gene can only increase the probability of descent of the same gene to relatives, and hence increase the chance of allelic identity. Correlations relative to a current pool of populations may be positive or negative.

We shall first (Section 4.2) extend the discussion of the previous chapter to encompass the regular mating systems. We shall present this analysis in terms of genotype frequency rather than gene-identity, but there is no essential difference (see (4.2.4)). Next (Section 4.3) we shall consider survival and extinction of genes and alleles in a finite population. Here again the reference point is some initial gene pool or set of founders, and the evolutionary process is defined by a population model, mating system or specified genealogy. In particular, we shall consider the effect of structural complexity on non-independence in gene-identity and gene extinction between different sets of genes. Thus, although we shall consider first the basepoint provided by the simple haploid models, it is also necessary to consider the evolution of higher order identity coefficients (Cockerham, 1971) and extinction of specified sets of genes in a set of diploid couples (Thompson, 1979*a*).

The next stage (Section 4.4) is an analysis of observable variability within and between current populations. The evolutionary model need not be specified; variability is partioned according to a (perhaps arbitrary) hierarchical structure. The theory of Wright (1921, 1951, 1965) is presented via the formulation of Cockerham (1969), and in addition to a general hierarchy we consider also mating systems and isolation by distance (Wright, 1943). To this point, our analysis has been of expectations, but, if analysis of variation is to be used as a basis for inference of sources of variation, this is not sufficient. Several statistics of current variation have been proposed as measures of evolutionary history (Cavalli-Sforza and Edwards, 1967; Nei, 1972; Morton *et al.*, 1971). In Section 4.5 we shall consider therefore the second moments of these statistics under simple evolutionary models. This provides a preliminary approach to problems of evolutionary inferences based on observable current populations, which we consider further in Chapter 5.

4.2   Loss of variability in regular mating systems

This section will extend the discussion of the preceding chapter to the problem of regular mating systems. Such systems are a further class of models of population structure, some simple systems being already covered by the models of Section 3.7(i) (sib-mating) and 3.9(iii) (double-first-cousin mating). The term *regular system* is normally applied to models in which all individuals in a generation have ancestors whose pattern of inter-relationship is the same, and in which the pattern of parent–offspring links is repeated at each generation. It may also be applied to systems in which the pattern of links is repeated, and the individuals of each generation have *as a group the same set* of patterns of inter-relationship amongst their ancestors, but in which not all individuals within a generation have the same ancestral relationships. An example is discussed by Thompson (1979*a*), but we shall consider only the simpler case of equivalent individuals. Of course, this does not imply that all pairs of individuals are equivalent. The essence of a mating system lies in the relationships *between* individuals who are chosen as mates.

Regular mating systems are considered in many standard texts. Although the notation here follows most closely that of Jacquard (1974), who also considers genotype frequencies, the discussion and classification of systems follows Kimura and Crow (1963). The regular systems are important in delimiting the possible patterns of inter-relationship between a group of individuals (Section 2.7) and the effects of consanguinous marriage. They also provide a simple class of structured systems for which other aspects of within-population loss of variability may be examined; we shall return to them again in Section 4.3 in discussing the extinction of genes.

For simplicity, consider a single autosomal locus carrying alleles $\{A_i: 1 \leq i \leq k\}$ with population frequencies $\{p_i: 1 \leq i \geq k\}$ in an infinite population. Since mating is restricted by genealogical relationship alone, a regular mating system cannot change these allele frequencies, but it does change the assortment of alleles within individuals (i.e. the genotype frequencies). Where all individuals derive via the same pattern of ancestral inter-relationship, it is sufficient for most purposes to consider the constructs

$$\boldsymbol{G}^{(t)} = \{G_{ij}^{(t)}, i \leq j\}, \qquad (4.2.1)$$

the array of frequencies (or probabilities) of genotypes A_iA_j at generation t. (Since we have an infinite population, frequencies and probabilities are equivalent.) Pairs of related individuals in this infinite population are not independent; to consider properties of pairs $\boldsymbol{G}^{(t)}$ is not sufficient.

Let us define also the *unit structures*

$$E_{ij} = E_{ji} \qquad (4.2.2)$$

having a 1 in the position corresponding to genotype $A_iA_j (i \leq j)$ and

0 elsewhere. Then a population in Hardy–Weinberg equilibrium has genotypic structure

$$G(\text{HW}) = \sum_i p_i^2 E_{ii} + \sum_{i<j}\sum \{2p_ip_jE_{ij}\} = \sum_i \sum_j p_ip_jE_{ij}. \qquad (4.2.3)$$

A population consisting only of homozygotes, but retaining allele frequencies p_i, has structure

$$G(I) = \sum_i p_iE_{ii}.$$

The evolution of the population under a mating system may be considered either in terms of probabilities of gene identity by descent, relative to an initial pool of individuals with Hardy–Weinberg frequencies, or directly in terms of genotype frequencies. The two are simply related. For suppose that the inbreeding coefficient (Section 2.2) of individuals at generation t is $\alpha^{(t)}(\alpha^{(0)} = 0)$, then

$$G^{(t)} = (1 - \alpha^{(t)})G(\text{HW}) + \alpha^{(t)}G(I). \qquad (4.2.4)$$

It is thus possible to consider $G^{(t)}$ under a variety of regular mating systems in terms of the simple structures $G(\text{HW})$ and $G(I)$ alone.

4.2(i)   *Sib-mating*

To introduce the methodology we shall consider sib-mating, although this is a special case of systems of maximum avoidance considered below. Under this system the population divides into lines of size 2, each couple having two offspring, a male and a female, who form a couple at the next generation (Figure 4.2(a)). Consider an individual B at generation t, and the origins of his two homologous genes in generation $(t-2)$. With probability $\frac{1}{2}$, these come from different individuals at $(t-2)$, and B is thus as if an

Figure 4.2(a). Sib-mating.

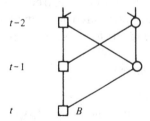

individual at generation $(t-1)$. With probability $\frac{1}{2}$, they come from the same individual, in which case there is a probability $\frac{1}{2}$ that B is an exact replicate of the $(t-2)$-generation individual, and $\frac{1}{2}$ that he receives two copies of the same gene. Thus

$$\boldsymbol{G}^{(t)} = \tfrac{1}{2}\boldsymbol{G}^{(t-1)} + \tfrac{1}{2}\{\tfrac{1}{2}\boldsymbol{G}^{(t-2)} + \tfrac{1}{2}\boldsymbol{G}(I)\}. \qquad (4.2.5)$$

Let

$$\Delta^{(t)} = \{\boldsymbol{G}^{(t)} - \boldsymbol{G}(I)\}2^{t},$$

then

$$\Delta^{(t)} = \Delta^{(t-1)} + \Delta^{(t-2)}$$

and $\Delta^{(t)}$ increases as the Fibonnacci series. Further $\boldsymbol{G}^{(t)} \to \boldsymbol{G}(I)$ at exponential rate λ, where λ is the largest root of the equation

$$\lambda^2 = \tfrac{1}{2}\lambda + \tfrac{1}{4}$$

or

$$\lambda = \tfrac{1}{4}(1+\sqrt{5})$$

(see Section 3.7(i)). Cockerham (1971) considers higher order identity coefficients under this mating system.

4.2(ii) *Parent–offspring mating*

In the case of matings between parent and offspring we have overlapping generations. The population again divides into lines of size 2, but it is convenient to label the generations as in Figure 4.2(b) with a single person in each generation. Consider again the origins of the genes of individual B at generation t in his grandparents F and M^*. With probability $\frac{1}{2}$, Bs genes come from F and M^* (via M not F). In this case B is equivalent to M, the child of F and M^*. With probability $\frac{1}{2}$, both Bs genes come from F, in which

Figure 4.2 (b). Parent–offspring mating.

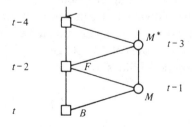

case, as in Section 4.2(i), B has probability $\frac{1}{2}$ of being a replicate of F, and $\frac{1}{2}$ of carrying identical genes. Thus

$$G^{(t)} = \tfrac{1}{2}G^{(t-1)} + \tfrac{1}{2}\{\tfrac{1}{2}G^{(t-2)} + \tfrac{1}{2}G(I)\}. \tag{4.2.6}$$

With this labelling of the generations, we have the same recurrence equation as before, and the same rate of approach of $G^{(t)}$ to $G(I)$.

4.2(iii) *Systems of maximum avoidance*

In subpopulations of size 2^m it is possible for matings to take place between dioecious individuals who have common ancestors no less than m generations ago. In this case all individuals are descended from all 2^m individuals of the population m generations previously; all matings are between individuals with 2^m common mth-generation ancestors. If an (infinite) population divides into such lines of size 2^m in which such a system is practised, we have a *system of maximum avoidance*. Special cases are selfing ($m = 0$), sib-mating ($m = 1$; see Section 4.2(i) above), double-first-cousin-mating ($m = 2$; see Section 3.8(iii)), quadruple-second-cousin-mating ($m = 3$), octuple-third-cousin-mating ($m = 4$), and so on. These systems (up to $m = 4$) were considered by Wright (1921).

Consider now the origins of the two homologous genes of individual B at generation t in his ancestors $(m + 1)$ generations earlier (cf. Section 4.2(i)). Each of these two genes (maternal and paternal) has independently a probability $(\frac{1}{2})^{m+1}$ of being each of these 2^{m+1} ancestral genes, since B's mother and father derive equally and independently from each of the 2^m ancestors at generation $(t - m - 1)$. Hence, with probability $(\frac{1}{2})^r$ $(1 \le r \le m)$, the genes derive from two distinct individuals at generation $(t - m - 1)$ who themselves have common ancestors r generations previously. The individual B will then have the genotype probability distribution of an individual at generation $(t - r)$. With the remaining probability $(\frac{1}{2})^m$, B's two genes derive from the same individual at generation $(t - m - 1)$, with probability $\frac{1}{2}$ causing B to be a replicate of that ancestor, and with probability $\frac{1}{2}$ being two replicates of a single gene.

Thus, combining the above possibilities, we have

$$G^{(t)} = \sum_{r=1}^{m} (\tfrac{1}{2})^r G^{(t-r)} + (\tfrac{1}{2})^m \{\tfrac{1}{2}G^{(t-m-1)} + \tfrac{1}{2}G(I)\}. \tag{4.2.7}$$

If we write $\Delta^{(t)} = \boldsymbol{G}^{(t)} - \boldsymbol{G}(I)$,

$$\Delta^{(t)} = \sum_{1}^{m+1} \{(\tfrac{1}{2})^r \Delta^{(t-r)}\}, \qquad (4.2.8)$$

which ensures convergence of $\Delta^{(t)}$ to zero since the sum of coefficients, $\sum_{1}^{m+1} (\tfrac{1}{2})^r = 1 - (1/2^{m+1})$, is strictly less than 1. The asymptotic rate of this convergence is the largest root of the equation

$$\lambda^{m+1} = \sum_{1}^{m+1} \{(\tfrac{1}{2})^r \lambda^{m+1-r}\}, \qquad (4.2.9)$$

which, for large m, has approximate solution

$$\lambda = 1 - 1/\{2^{m+2} - (m+1)\} \qquad (4.2.10)$$

(Robertson, 1964).

The *effective population size* in diploids is thus approximately $\{2^{m+1} - \tfrac{1}{2}(m+1)\}$ or of the order of twice the actual size, $N = 2^m$. The approximation

$$(1-\lambda) \simeq 1/2^{m+2} \simeq 1/4N \qquad (4.2.11)$$

was given earlier by Wright (1933).

4.2(iv) Half-sib mating

There are several different ways in which repeated matings between half-sibs may be constructed. One system, most appropriate in the case of animal breeding, is where a single male

Figure 4.2(c). Half-sib mating; unlimited population.

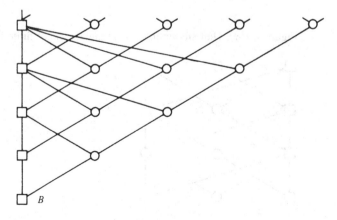

produces a male, and a large number of females, all by different females. In the next generation each of these females is mated to the male, producing one male and a large number of females for the subsequent generation (Figure 4.2(c)). This system was considered by Wright (1921) and the genotypic recursions are given by Jacquard (1974), but the system has problems as a population model, since the lines are not of constant size. Wright (1921) therefore considered an alternative (Figure 4.2(d)). In this system every generation consists of a male and two females, the females being full sibs, but only half-sibs of the male.

A more mathematically satisfying alternative is the case of m males and m females mated cyclically (Figure 4.2(e)). In this case all individuals within a generation are equivalent, and each produces two offspring, a male and a female. We cannot consider a recurrence relation only in terms of $\boldsymbol{G}^{(t)}$, since genes in B at generation t may derive from individuals separated by r steps [mod $2m$] at generation $(t-r)$, but all matings are between adjacent individuals. To circumvent this problem we define

$$\{\boldsymbol{G}_i^{(t)}: 0 \le i \le m\} \tag{4.2.12}$$

to be the array of genotype frequencies that *would* result at generation t if half-sib mating were practised up to generation $(t-1)$, but then individuals i steps apart were mated. Then, since actual matings are between adjacent individuals, we have (see Figure 4.2(e)),

$$\boldsymbol{G}_i^{(t)} = \tfrac{1}{2}\boldsymbol{G}_i^{(t-1)} + \tfrac{1}{4}\boldsymbol{G}_{(i-1)}^{(t-1)} + \tfrac{1}{4}\boldsymbol{G}_{(i+1)}^{(t-1)}, \qquad 1 \le i \le m \tag{4.2.13}$$

Figure 4.2(d). Half-sib mating: single male; case of finite lines.

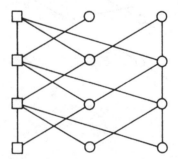

where

$$G_i^{(t)} = G_{-i}^{(t)} = G_{2m-i}^{(t)} \qquad (4.2.14)$$

all subscripts being defined modulo $2m$. In addition we have

$$G_0^{(t)} = \tfrac{1}{2}G_1^{(t-2)} + \tfrac{1}{2}G(I), \qquad (4.2.15)$$

since $G_0^{(t)}$ is the genotypic structure of a pair of genes chosen independently from the same individual at generation $(t-1)$. Thus with probability $\tfrac{1}{2}$ the same gene is chosen twice, giving $G(I)$, and with probability $\tfrac{1}{2}$ the $(t-1)$-generation individual is replicated, this individual being the result of mating adjacent individuals at generation $(t-2)$.

Equations (4.2.13)–(4.2.15), together with initial conditions $G_i^{(0)} = G(\mathrm{HW})$, $1 \le i \le m$, are sufficient to determine $G_i^{(t)}$ for all i and t, and in particular $G_1^{(t)}$ which is the required result for half-sib mating. (An equivalent set of equations is given by Kimura and Crow (1963).) If we again write

$$\Delta_i^{(t)} = G_i^{(t)} - G(I)$$

we have

and

$$\left. \begin{aligned} \Delta_i^{(t)} &= \tfrac{1}{4}\{\Delta_{i-1}^{(t-1)} + 2\Delta_i^{(t-1)} + \Delta_{i+1}^{(t-1)}\}, \ 1 \le i \le m \\ \Delta_0^{(t)} &= \tfrac{1}{2}\Delta_1^{(t-2)}. \end{aligned} \right\} \qquad (4.2.16)$$

Thus $\Delta_i^{(t)} \to 0$ for all i, and the asymptotic rate of approach is given by the largest eigenvalue of the system of equations (4.2.16). The rate is clearly the same for $i=0$ as for all $i \ne 0$; the rate of loss of heterozygotes is the same as the rate of loss of heterozygosity in the group. Kimura and Crow (1963) show this largest eigenvalue to be approximately

$$\begin{aligned} \lambda &= 1 - \pi^2/16(m+1)^2 \\ &\simeq 1 - \pi^2/(2N)^2 \quad (N = 2m). \end{aligned} \qquad (4.2.17)$$

Thus the asymptotic approach to $G(I)$ is extremely slow, the

Figure 4.2(*e*). Cyclic half-sib mating.

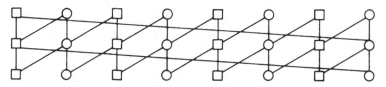

effective population size being of the order of the square of the number of individuals. The rate is in fact slower than for systems of *maximum avoidance*. As discussed by Robertson (1964), the maximum avoidance schemes maintain the lowest inbreeding rates in the early generations, but this is at the expense of a higher asymptotic rate of convergence to $G(I)$.

4.2(v) *First-cousin mating*

For all the above systems $G^{(t)} \to G(I)$. This is necessarily the case where all ancestors of an individual are ultimately related, and hence the case where the population divides into finite lines. The limiting case is that of first-cousin mating, where any individual possesses a pair of mth-generation ancestors who are only mth-cousins (Figure 4.2(f)). Although there are pairs of ancestors whose coefficient of kinship is arbitrarily small, there are no unrelated pairs, and thus $G^{(t)} \to G(I)$. That $G(I)$ is the only equilibrium for this system was shown by Wright (1921), while Jacquard (1974) gives the recurrence equations enabling the genotype frequencies at all generations to be found. Again we cannot consider directly a recursion for $G^{(t)}$, since an individual has ancestors who are cousins of arbitrary degree. Analogously to Section 4.2(iv) we instead define $G_i^{(t)}$ to be the genotype structure formed by performing first-cousin mating to generation $(t-1)$, but then pairing ith cousins. Then, again as in Section 4.2(iv) we have

$$G_i^{(t)} = \tfrac{1}{4}\{G_{i-1}^{(t-1)} + 2G_i^{(t-1)} + G_{i+1}^{(t-1)}\}, \qquad 1 \le i \le \tfrac{1}{2}t \qquad (4.2.18)$$

Figure 4.2(f). First-cousin mating.

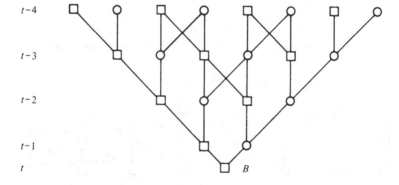

while considering the offspring of sibs (0th-cousins) at generation $(t-1)$,

$$\boldsymbol{G}_0^{(t)} = \tfrac{1}{4}\{\boldsymbol{G}(I) + 2\boldsymbol{G}_1^{(t-1)} + \boldsymbol{G}_1^{(t-2)}\}. \tag{4.2.19}$$

This last equation holds since the genes taken from sibs at $(t-1)$ are, with respect to generation $(t-3)$, the same gene with probability $\tfrac{1}{4}$, come from first-cousins at $(t-2)$ with probability $\tfrac{1}{2}$ (giving $\boldsymbol{G}_1^{(t-1)}$), and from first-cousins at $(t-3)$ with probability $\tfrac{1}{4}$. Note the parallel between (4.2.18), (4.2.19) and (4.2.13), (4.2.15). Whereas half-sib mating is mating of adjacent individuals, first-cousin mating is mating of offspring of adjacent couples. From (4.2.18) and (4.2.19), with initial Hardy–Weinberg equilibrium, the convergence of $\boldsymbol{G}_1^{(t)}$ (and indeed $\boldsymbol{G}_i^{(t)}$ for all i) to $\boldsymbol{G}(I)$ can be established, but this convergence is not exponential.

Kimura and Crow (1963) consider a modification of the above system, where first-cousin mating is achieved in a population of constant size $2m$ by a *circular pair mating* system. Individuals $2m$ steps apart in a given generation of Figure 4.2(f) are superposed, and the indices of (4.2.18) must now be evaluated modulo $2m$. In this finite system we necessarily have exponential convergence to $\boldsymbol{G}(I)$. Kimura and Crow showed that the rate of this convergence is

$$\lambda = 1 - \{\pi^2/4(m+6)^2\} \quad \text{(for large } m\text{)}.$$

Thus circular first-cousin mating gives the same rate of convergence to $\boldsymbol{G}(I)$ (for large m) as half-sib mating in a population half the size. This is perhaps expected in view of the above comment that cyclic first-cousin mating is analogous to cyclic half-sib *mating of couples*.

4.2(vi) *More remote relationships*

For systems of mating between individuals more remotely related than first-cousins, an individual has unrelated ancestors. (This is of course only possible in an infinite population.) The two cases of half-first-cousin mating and second-cousin mating were considered briefly by Wright (1921), and Jacquard (1974) gives recurrence equations for the genotypic frequencies for the latter case. The main interest of such systems lies, however, in the equilibrium, where not all heterozygotes are lost, rather than in the approach to it, which is the main feature of systems whose only equilibrium is $\boldsymbol{G}(I)$. We therefore consider only the equilibrium, which may be quite simply derived.

In the case of matings between mth-cousins ($m \geq 2$) consider the origin of genes of individual B in his 2^m paternal and 2^m maternal ancestors ($m + 1$) generations previously. There is a probability $(\frac{1}{2})^m$ that his paternal gene derives from any particular paternal ancestor, and the descent of paternal and maternal genes are independent. Now one maternal–paternal pair of ancestors are sibs, since B's parents are mth-cousins. In addition, one other maternal and one other paternal ancestor is each an mth-cousin of this pair of sibs (but unrelated to each other) since these are the mates of this sib pair (Figure 4.2(g)). However, all other maternal–paternal pairs are unrelated, although of course there are many inter-relationships within the maternal and paternal sets. Thus there is a probability $(\frac{1}{2})^{2m}$ that the genes of B derive from a pair of sibs, and $2/2^{2m}$ that they derive from mth-cousins, and hence

$$\{1 - 3/2^{2m}\} \tag{4.2.20}$$

that they derive from an unrelated pair. Further, if the genes derive from the sib pair, there is a probability $\frac{1}{4}$ that they are identical owing to this sib relationship. Thus the inhomogeneous term in the genotypic recurrence relation is

$$\left\{1 - \frac{3}{2^{2m}}\right\} \boldsymbol{G}(\mathrm{HW}) + \frac{1}{4}\frac{1}{2^{2m}} \boldsymbol{G}(I) \tag{4.2.21}$$

Figure 4.2(g). mth-cousin mating ($m = 2$).

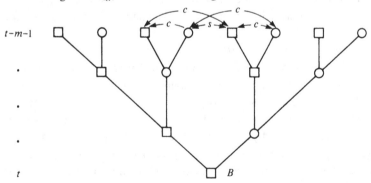

s = sibs

c = mth-cousins

and convergence to an equilibrium structure

$$\{4(2^{2m}-3)\boldsymbol{G}(\text{HW})+\boldsymbol{G}(I)\}/\{4(2^{2m}-3)+1\} \tag{4.2.22}$$

is assured.

In the case of $\frac{1}{2}$-mth-cousin mating ($m \geq 1$) it is more convenient to consider the origin of Bs genes ($m+2$) generations previously. Again, descents of maternal and paternal genes to B from this generation are independent, although one of the 2^{m+1} maternal ancestors is identical to one of the 2^{m+1} paternal ancestors. That is, there are in fact only ($2^{m+2}-1$) ancestors (Figure 4.2(h)). Another maternal and another paternal ancestors are each mates of the duplicated ancestor, and hence her $\frac{1}{2}$-mth-cousin, although again the two are unrelated to each other. All other maternal–paternal pairs are also unrelated. Thus there is now a probability

$$\{1-3(\tfrac{1}{2})^{2(m+1)}\} \tag{4.2.23}$$

(cf. (4.2.20)) that Bs genes derive from unrelated ancestors, in the ($m+2$)-nd preceding generation, and $(\tfrac{1}{2})^{2(m+1)}$ that both genes derive from the same individual, and hence

$$\frac{1}{2}\frac{1}{2^{2(m+1)}}$$

that he receives the same gene twice over. Thus the inhomogeneous term of the recurrence is

$$\left(1-\frac{3}{2^{2(m+1)}}\right)\boldsymbol{G}(\text{HW})+\frac{1}{2^{2m+3}}\boldsymbol{G}(I), \tag{4.2.24}$$

and we have convergence to the equilibrium

$$\{2(2^{2(m+1)}-3)\boldsymbol{G}(\text{HW})+\boldsymbol{G}(I)\}/\{2(2^{2(m+1)}-3)+1\}. \tag{4.2.25}$$

Figure 4.2(h). $\frac{1}{2}$-mth-cousin mating ($m=1$).

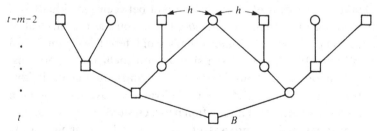

h = half – mth – cousins

The main point of interest is how little the equilibria (4.2.22) and (4.2.25) differ from the initial Hardy–Weinberg frequencies, even for small m.

For $m = 2$, (4.2.22) becomes $\frac{1}{53}\{52\boldsymbol{G}(\text{HW}) + \boldsymbol{G}(I)\}$.

For $m = 1$, (4.2.25) becomes $\frac{1}{27}\{26\boldsymbol{G}(\text{HW}) + \boldsymbol{G}(I)\}$.

The excesses of homozygotes are only $\frac{1}{53}$ and $\frac{1}{27}$ respectively. Thus as soon as we move from systems where all variability is lost, to those where some ancestors are unrelated, genotype frequencies are approximately as in an infinite random mating population, even though all matings are between related individuals. 'Small amounts' of heterozygosity cannot be maintained (in the limit) by a regular mating system.

4.3   Gene survival and mating structure

Robertson (1964) notes that the higher the rate of inbreeding within a line the smaller the ultimate rate of loss of variability, provided there is no permanent division of lines (see Section 4.2(iv)). In an infinite population the original Hardy–Weinberg equilibrium is of course immediately restored by a single generation of random mating. Even in a finite population the ultimate rate of loss may be made arbitrarily small by interspersing random mating at long intervals on a high-inbreeding system such as sib-mating. The loss of alleles is slower than in a random mating population. The ultimate rate of loss of variability is thus not the major factor of practical importance in characterizing a mating system, and in this section we consider an alternative feature – the short-term survival and extinction of genes.

Of course, gene extinction and loss of variability are closely related aspects of the same process. There is, however, a major difference in the direction from which we view the process. The identity by descent of genes within and between individuals is a function of their common *ancestry*, but in considering survival of ancestral genes it is *descendants* that are of interest. As commented by Wright (1965), in his discussion of path coefficient algorithms, '. . . common descendants do not tend to contribute to correlations between ancestors'. Yet where individuals have no common descendants there is no interaction between survival of their genes. We have studied the process of loss of variability or increasing gene-identity by starting at a Hardy–Weinberg situation and considering the frequencies $G^{(t)}$ that will obtain t generations *later*. In analysing gene extinction, we consider a monomorphic endpoint, and to compute the probabilities that certain genotype combinations t generations *earlier* will give rise to this endpoint (Thompson, 1979a).

4.3(i)   *Rate of loss of multiple alleles*

The large-population diffusion approximation to the random mating Wright model (Crow and Kimura, 1970) has provided several results concerning loss of alleles from a (large) constant-sized finite population of N haploid individuals. Kimura (1955) considered the rate of loss of alleles in the triallelic case, and Littler (1975) extended this to arbitrary numbers of alleles. Griffiths (1980), as a special case of a more general problem, considers the number of genes, $g(t)$ say, surviving after Nt generations. He gives the form

$$Q_l(t) = p(g(t) = l)$$

$$= \sum_{r=l}^{\infty} \left\{ e^{-\frac{1}{2}r(r-1)t}(-1)^{r-1}\binom{r}{l} l_{(r-1)} \frac{(2r-1)}{r!} \right\} \qquad (4.3.1)$$

where $l_{(r-1)} = (l+1)(l+2)\ldots(l+r-2)$. The expected number of surviving genes is thus

$$E(g(t)) = \sum l Q_l(t) = \sum_{r=1}^{\infty} (2r-1) e^{-\frac{1}{2}r(r-1)t}. \qquad (4.3.2)$$

Further, let T_l denote the time (in units of N generations) until only l genes survive:

$$P(T_l > t) = P(g(t) \geq l+1),$$

and

$$E(T_{l-1} - T_l) = \int_0^{\infty} \{P(T_{l-1} > t) - P(T_l > t)\} \, dt$$

$$= \int_0^{\infty} P(g(t) = l) \, dt = \int_0^{\infty} Q_l(t) \, dt$$

$$= 2/l(l-1). \qquad (4.3.3)$$

Thus, when l genes survive, the rate of loss of genes is $\frac{1}{2}l(l-1)$, as given by Littler (1975). Note that (4.3.1) provides also for the elimination of alleles with any specified initial frequencies. These frequencies determine the probability A_{ij} that a sample of i initial genes contains j alleles. Then if $a(t)$ is the number of surviving alleles at t

$$P(a(t) = k) = \sum_{j=k}^{\infty} A_{jk} P(g(t) = j). \qquad (4.3.4)$$

The same results are provided by the more general formulation of Karlin (1968c) and Felsenstein (1971a), without using the diffusion approximation. The basic entities are the probabilities Γ_{ij} that i genes derive from j parental genes (Section 3.5). Then clearly

$$A_{ij}^{(t)} = \sum_{k=1}^{N} \Gamma_{ik} A_{kj}^{(t-1)}, \qquad (4.3.5)$$

where $A_{ij}^{(t)}$ is now the probability that i genes at generation t contain j alleles. Felsenstein (1971a) shows that the rate of loss of alleles, when j remain, is $(1 - \Gamma_{jj})$, which for the haploid Wright model is

$$1 - \prod_{r=1}^{(j-1)} \left(1 - \frac{r}{N}\right), \qquad (4.3.6)$$

since the probability of all j individuals having distinct parents is the latter product. If we consider time in units of N generations, and let $N \to \infty$, we have

$$\Gamma_{jj}^{(N)} = (\Gamma_{jj})^N \to e^{-\frac{1}{2}j(j-1)},$$

or the rate of loss (4.3.3) obtained above.

The quantities of interest are $A_{Nj}^{(t)}$, the probabilities of j surviving alleles in the whole population. These are obtainable from (4.3.5), but it is often simpler to consider first the loss of genes, and proceed to alleles via (4.3.4). If we consider now each gene at $t = 0$ to be of a distinct allelic type, $A_{ij}^{(0)}$ is the identity matrix and

$$\mathbf{A}^{(t)} = \mathbf{\Gamma} \mathbf{A}^{(t-1)} = \mathbf{\Gamma}^t = \mathbf{A}^{(t-1)} \mathbf{\Gamma}$$

or

$$A_{ij}^{(t)} = \sum_{k=1}^{N} A_{ik}^{(t-1)} \Gamma_{kj}.$$

This form has the advantage that to obtain the probabilities of j surviving genes we need consider only $\{a_j^{(t)} = A_{Nj}^{(t)}; 1 \le j \le N\}$ rather than the complete matrix $\mathbf{A}^{(t)}$ required for (4.3.5). We have

$$a_j^{(t)} = \sum_{k=1}^{N} a_k^{(t-1)} \Gamma_{kj}. \qquad (4.3.7)$$

Note that the factor Γ now refers to transitions in the earliest generation considered. We shall find it convenient to approach gene extinction from a backwards viewpoint. Rather than considering evolution from a zero origin to a point t generations later, we shall consider survival from points t generations ago, to the present (*zero*). The form (4.3.7) is therefore appropriate, and it is not necessary for population size to remain constant; $a_k^{(t-1)}$ is the probability of survival of precisely k genes from $(t-1)$ generations ago, and Γ_{kj} is the probability of k genes at $(t-1)$ generations ago deriving from j genes at t generations ago. For the Wright model Γ_{ij} is the number of ways i items can be assigned to N boxes (potential parents) in such a way that precisely j are occupied; that is

$$\Gamma_{ij} = \binom{N}{j} \sum_{k=0}^{(j-1)} (-1)^k \binom{j}{k} \left(\frac{j-k}{N}\right)^i, \qquad 1 \le i \le N^* \qquad (4.3.8)$$

(Feller, 1968, p. 60), where N and N^* are population sizes t and $(t-1)$ generations ago.

4.3(ii) *Gene extinction*

The results of Felsenstein (1971a) are applicable to any model without mutation and selection, in which all individuals are equivalent with regard to reproduction, and in which the model does not vary over time. In particular, they are applicable to the conditioned branching process models of Karlin and McGregor (1964). We shall consider these models with a view to obtaining a more complete characterization than the asymptotic loss rates. Of particular interest will be the extent to which extinction of some set of genes affects extinction probabilities of others. We therefore require extinction probabilities of arbitrary specified sets of ancestral genes. To achieve this, consider a diallelic locus. If the genes of interest are labelled as alleles A_1, and the remainder A_2, then the probability of extinction of (at least) the set of interest over t generations is the probability of loss of the allele A_1 over that period.

For a model in which all N haploid individuals are interchangeable it is sufficient therefore, to consider $P_{ij}^{(t)}$, the probability of transition from i to j copies of an allele over t generations. Further, let

$$q_k^{(t)} = P_{k0}^{(t)}$$

be the extinction probability of any specified set of k genes. The probabilities $q_k^{(t)}$ are, of course, related to the $a_k^{(t)}$ above. There are in all $\binom{N}{j}$ sets of size j, each having extinction probability $a_{N-j}^{(t)}/\binom{N}{j}$. $\binom{N-k}{j-k}$ of these subsets contain a specified set of k genes:

$$q_k^{(t)} = \sum_{j \geq k} \left\{ \binom{N-k}{j-k} a_{N-j}^{(t)} \Big/ \binom{N}{j} \right\}. \qquad (4.3.9)$$

However (4.2.9) is too complicated to be useful: it is simpler to consider separately the $a_j^{(t)}$, via (4.3.7), and the $q_j^{(t)}$ via

$$q_j^{(t)} = \sum_{k=0}^{N} P_{jk} q_k^{(t-1)}, \qquad \text{where } P_{jk} = P_{jk}^{(1)}. \qquad (4.3.10)$$

Note that the transition from j to k copies of allele A_1 in (4.3.10) again refers to transition in the earliest generation. Thus, with this proviso, (4.3.10) is applicable to varying population size. For a Wright model the analogue of (4.3.8) is

$$P_{jk} = \binom{N^*}{k} \left(\frac{j}{N}\right)^k \left(1 - \frac{j}{N}\right)^{(N^*-k)}. \qquad \begin{matrix} 0 \leq j \leq N \\ 0 \leq k \leq N^* \end{matrix} \qquad (4.3.11)$$

In a conditioned branching process model, (haploid) individuals replicate independently, but the result is constrained to consist of precisely N (a constant) offspring. The Wright model is obtained in the case of Poisson progeny distribution. Let $g(z) = \sum_0^\infty g_r z^r$, $(g_r \geq 0$, $\Sigma g_r = 1)$ be the generating function of the progeny distribution, g' its derivative, and $C(z': f(z))$ denote the coefficient of z^r in the power series expansion of $f(z)$ in non-negative powers of z. Then

$$P_{jk} = C(z_1^k z_2^{N-k} : (g(z_1))^j (g(z_2))^{N-j}) / C(z^N : (g(z))^N). \qquad (4.3.12)$$

Further the eigenvalues of $\boldsymbol{P}$ are $\lambda_0 = 1$,

$$\lambda_r = C(z^{N-r} : (g(z))^{N-r} (g'(z))^r) / C(z^N : (g(z))^N) \qquad (1 \leq r \leq N) \qquad (4.3.13)$$

Karlin (1966) shows further that these eigenvalues are in decreasing order, and

$$\sum_{k=0}^{N} P_{jk} k^r = \lambda_r j^r + u_{r-1}(j), \qquad (4.3.14)$$

where $u_{r-1}(j)$ is a polynomial of degree $\leq (r-1)$ in j.

Now setting $r = 0$ in (4.3.14), $\sum_0^N P_{jk} = 1$, $\lambda_0 = 1$, and the first right eigenvector

$$\boldsymbol{x}_0 = (1, 1, \ldots, 1)'$$

When $r = 1$ we have

$$\sum_0^N P_{jk}k = E(A_1 \text{ genes in next generation}|\text{now } j)$$

$$= j, \text{ for a population of constant size.}$$

Thus $\lambda_1 = 1$ and

$$\mathbf{x}_1 = (x_1(i)) = (i; 0 \le i \le N).$$

When eigenvectors $\mathbf{x}_0,, \ldots, \mathbf{x}_r$ and corresponding eigenvalues $\lambda_0, \ldots, \lambda_r$, are determined, assume $\mathbf{x}_{r+1} = (\sum_0^{r+1} b_j i^j; 0 \le i \le N)$. Then

$$\lambda_{r+1}\left\{\sum_{l=0}^{r+1} b_l j^l\right\} = \sum_0^N P_{jk}\left\{\sum_{l=0}^{r+1} b_l k^l\right\}$$

$$= \sum_{l=0}^{r+1} b_l\{\lambda_l j^l + u_{l-1}(j)\},$$

from (4.3.14). Since the previous eigenvalues are known, λ_{r+1} and the coefficients $\{b_l\}$ may now be determined by equating powers of j. In the case of the Wright model

$$\left.\begin{aligned} P_{jk} &= \binom{N}{k}j^k(N-j)^{N-k}/N^N \\[2mm] \lambda_r &= \prod_{l=1}^{(r-1)}\left(1-\frac{l}{N}\right) \end{aligned}\right\} \tag{4.3.15}$$

(Feller, 1951, cf. (4.3.6)), and

$$\left.\begin{aligned} \mathbf{x}_2 &= (x_2(i) = i(i-N); 0 \le i \le N) \\ \mathbf{x}_3 &= (x_3(i) = i(i-N)(i-\tfrac{1}{2}N); 0 \le i \le N). \end{aligned}\right\}$$

Piva and Holgate (1977) have given the eigenvectors in closed form, but the recursive procedure of Karlin (1966) is more readily applicable in practice.

Right eigenvectors $\mathbf{y}_r = (y_r(j); 0 \le j \le N)$, orthonormal to $\{\mathbf{x}_r\}$, may also be determined recursively, but in this case those corresponding to the smallest non-unit eigenvalues, $\lambda_N, \lambda_{N-1} \ldots$ are the first computed. For general results the procedure is therefore not useful, but for numerical results in small populations it provides a simple method of computing extinction probabilities:

$$\mathbf{P}^t = \sum_{r=0}^N \lambda_r^t \mathbf{x}_r \mathbf{y}_r'$$

and

$$q_k^{(t)} = (\boldsymbol{P}^t)_{k0} = \sum_{r=0}^{N} \lambda_r^t x_r(k) y_r(0). \tag{4.3.16}$$

Thus knowledge of $y_r(0)$ is sufficient to determine extinction probabilities, the case $r = 2$ providing the dominant asymptotic behaviour, since $\lambda_2 < 1 = \lambda_1 = \lambda_0$. For the Wright model again

$$q_k^{(t)} \simeq \left(1 - \frac{k}{N}\right) - \left(1 - \frac{1}{N}\right)^t k(N-k) y_2(0),$$
$$\left(y_0(0) = \frac{1}{N}, \ y_1(0) = -\frac{1}{N}\right). \tag{4.3.17}$$

4.3(iii) *Correlations and structure*

For a model of reproductively equivalent haploids, the only correlations are those between a specified set of r genes and a disjoint specified set of s genes:

$$\rho_{r,s}^{(t)} = \frac{\{q_{r+s}^{(t)} - q_r^{(t)} q_s^{(t)}\}}{\{q_r^{(t)}(1-q_r^{(t)}) q_s^{(t)}(1-q_s^{(t)})\}^{1/2}}, \tag{4.3.18}$$

which, from (4.3.17), behaves asymptotically for the Wright model as

$$\rho_{r,s}^{(t)} \simeq -\left\{\frac{rs}{(N-r)(N-s)}\right\}^{1/2} \left\{1 - N^2 y_2(0)\left(1 - \frac{1}{N}\right)^t\right\}.$$

For all r and s $(r+s < N)$ the absolute value of the correlation increases monotonically in time to its limiting value. All correlations are negative; extinction of specified genes increases the survival chances of others. These haploid correlations provide a basepoint for comparison with those arising in structured systems.

One aspect of structure is changing population size, which can have a major impact on mutant survival (Fisher, 1930) and hence on observed variants within a population (Thompson and Neel, 1978). The size factor having the largest single effect on population variability is population bottlenecks (Nei, 1973). The effective population size under varying actual size is the harmonic mean of these (section 3.2.(vi)); the small values are the critical ones. The effect on correlations in extinction of disjoint sets of genes is more complex; the pattern of consecutive sizes is also relevant (see (4.3.11)). These effects have been investigated numerically via (4.3.11) and (4.3.18), but there are no explicit mathematical results.

Within a population, the lowest level of structure is provided by the existence of diploid individuals. In the random mating case, correlations between the zero/one variables of survival/extinction of disjoint sets of ancestral genes depend on the initial distribution of the genes amongst individuals; a pair of genes is initially within the same individual or in different ones. Rather than the number of A_1 genes in the population, we must now consider numbers of the three genotypes A_1A_1, A_1A_2, A_2A_2. Again, labelling the genes of interest as A_1, extinction of these genes corresponds to transition to the state of $N\, A_2A_2$ individuals. Numerical results for small N may therefore be obtained. Thompson (1979a) discussed the case $N = 4$).

There are two further aspects of structure. One is given by a more complex offspring distribution (Section 3.2(ii)) and the other by restrictions on the mating patterns (Section 3.9). Since interactions in extinction of genes of ancestors arise owing to Mendelian segregation in their common descendants, a fundamental component is the correlation in extinction of genes of a couple, by virtue of their common children. For the basic case we assume these children to have no descendants in common, extinction of their genes being therefore independent. The couple is not part of the *current* population; the children may be. Denote the genes of the couple a, b, c, d and those of the children g_i $(1 \le i \le 2k)$ (Figure 4.3(a)). Let $q(S)$ denote the extinction probability of a set of genes S, over a specified section of pedigree. Then

$$
q(a, b) = q(c, d) = \prod_{j=1}^{k} q(g_{2j-1}, g_{2j})
$$

$$
q(a, b, c) = q(a, c, d) = \prod_{j=1}^{k} \{\tfrac{1}{2}(q(g_{2j-1}, g_{2j}) + q(g_{2j}))\}
$$

$$
q(a, c) = \prod_{j=1}^{k} \{\tfrac{1}{4}(q(g_{2j-1}, g_{2j}) + 2q(g_{2j}) + 1)\} \qquad (4.3.19)
$$

and

$$
q(a) = q(c) = \prod_{j=1}^{k} \{\tfrac{1}{2}(1 + q(g_{2j}))\} \qquad (*)
$$

(note $q(g_{2j}) = q(g_{2j-1})$),

since with probability $\tfrac{1}{2}$ any specified parental gene is passed to a given child, and the two genes within any individual are equivalent.

Equation (4.3.19) provides the basic components required for correlations between genes of a couple: for example

$$\rho(\{a, b\}, \{c\}) = \frac{q(a, b, c) - q(a, b)q(c)}{\{q(a, b)(1 - q(a, b))q(c)(1 - q(c))\}^{1/2}}.$$

In order to consider correlations between ancestors of the couple, resulting from Mendelian segregation in their offspring, we require only the additional factor of the probabilities of subsets of genes a, b, c, d deriving from these ancestors. Next we may consider the effect of several such descendant couples, and finally the effect of other routes of non-interacting descent from the original ancestors (Figure 4.3(a)). In each case the basic components of correlation are provided by (4.3.19). The absolute magnitude of correlations decreases with numbers of offspring, numbers of spouses, increasing number of generations from interacting couples and increasing alternative routes of descent. All correlations induced by a specific pedigree structure are negative, although a model in which family

Figure 4.3(a). Patterns of descent from two sets of ancestral genes.

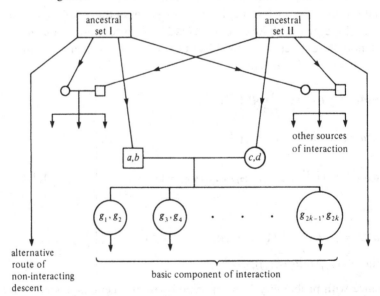

size distribution is a random variable can provide positive correlations. For example, if there is a non-zero probability of no
offspring, extinction of genes of an individual increases the probability of this event, and hence *increases* extinction probability of
genes of spouses.

Equation (4.3.19*) also provides an extremely rapid recursive
method for computing extinction probabilities on an outbreeding
pedigree. Further, if $h_g(z)$ denotes the generating function of the
current number of replicates of an ancestral gene, g, we have, as in
(4.3.19*)

$$h_a(z) = z^{\delta(a)} \prod_{j=1}^{k} \tfrac{1}{2}(1 + h_{g_{2j}}(z))$$

where $\delta(a) = 0/1$ as a is not/is part of the current population. Thus
patterns of gene survival in outbreeding populations are particularly simple to analyse: this case is discussed further by Thompson
(1979*a*).

To analyse more structured populations, the regular mating
systems (Section 4.2) provide a useful basepoint. We again require
transition probabilities from each distinct genotype pattern at a
diallelic autosomal locus, to complete monomorphism, over t
generations. Where genotype transition probabilities can be
explicitly determined, as in the case of sib-mating (Feller, 1968),
extinction probabilities of arbitrary sets of genes and hence
also correlations in extinction follow. In other cases the method
of Cannings, Thompson and Skolnick (1978) is applicable. We
again work backwards from a *current* monomorphic A_2A_2 population, determining recursively $(t = 1, 2, 3 \ldots)$ the transition
probabilities from each possible genotype combination t generations earlier, to this endpoint. This may be done simultaneously for
all initial genotype combinations (Thompson, Cannings and Skolnick, 1978). Thompson (1979*a*) has considered several such mating
systems, and discussed the results.

Whatever aspect of a system or population model is considered, it
is of course a function of the transition matrix of genes Γ (4.3.5) or
allelic copies P (4.3.12), these matrices being themselves closely
related (4.3.9). Where a system has complex structure, more
detailed population states must be specified, but an explicit form of
P^t or G^t must still provide the complete solution. Asymptotically at

least the leading eigenvalues considered in Chapter 3 are thus a sufficient characterization. However, where a full solution is not available, consideration of gene extinction, and short-term interactions in gene extinction, does provide further characterization. Further, this characterization is not directly analogous to that provided by the genotype transitions of Section 4.2. Unfortunately, in many cases, only numerical results are possible, and we have therefore presented here only the general methodology and outline of the variety of structural patterns that can be considered via this methodology.

4.4 Partition of genetic variability

When a population, or group of populations, is observed, we may see variation in allele frequency and/or deviations in the genotype frequencies from Hardy–Weinberg equilibrium. Normally we shall know only the current frequencies, and often the historical structure of the population will be unavailable. Nonetheless we shall wish to provide some analysis of the genetic variability. The aim of this section is to provide a view of this variability, in terms of a hierarchic structure of correlations, without any underlying assumptions about its source.

4.4(i) *Partition of variance*

The simplest case is that of an infinite population (hence constant allele frequencies), in which consanguinous matings occur. Thus the pair of genes within an individual are not a randomly chosen pair from the population; our hierarchic structure has two levels – genes within and between individuals. This departure from randomness is of course manifested by a departure from Hardy–Weinberg genotype frequencies. Consider an autosomal locus with alleles A_i with population frequencies p_i. Suppose that the probability that autosomal homologous genes uniting in an individual are identical by descent relative to the founders of the population is α (see (2.2.3)). Then the frequency of A_iA_i individuals will be

$$P_{ii} = p_i^2(1-\alpha) + p_i\alpha$$
$$= p_i^2 + \alpha p_i(1-p_i), \tag{4.4.1}$$

since an individual carrying identical genes is necessarily homozygous (in the absence of mutation), and there is probability p_i that the allele carried is A_i, whereas, if he carries distinct genes, there is probability p_i^2 that both are A_i. Similarly

$$P_{ij} = 2p_1p_j(1 - \alpha), \quad \text{for } i \neq j. \tag{4.4.2}$$

The parameter α is the *fixation index* of Wright (1921). Although in this simplest case this index is interpretable as a probability of identity by descent, in fact the inbreeding coefficient (2.2.3), we see from (4.4.1) that it is equally the correlation between two 0/1 random variables defined for the pair of genes;

$$X = \begin{cases} 1 & \text{if gene is allelic type } A_i \\ 0 & \text{otherwise.} \end{cases} \tag{4.4.3}$$

It is this latter aspect of correlations that we shall pursue in this section. In addition to fixation indices for neutral autosomal genes, there are also indices appropriate to sex-linked or polysomic loci (Wright, 1951). We shall, however, consider only autosomal loci.

Now consider a particular allelic type (A_1 say), grouping the remainder together as A_2 with initial frequency $p_2 = 1 - p_1$. Suppose we have a finite collection of N diploid individuals. Following Cockerham (1969) write

$$X_{ij} = \begin{cases} 1 & \text{if } j\text{th gene of } i\text{th individual is } A_1 \\ 0 & \text{otherwise} \end{cases}$$

$$(j = 1, 2, 1 \leq i \leq n) \tag{4.4.4}$$

Then the allele frequency observed in the population is

$$\bar{X}_{..} = \left(\frac{1}{2N}\right) \sum_i \sum_j X_{ij},$$

where the dots denote subscripts over which means have been taken, while, relative to some foundation stock with allele frequency p_1,

$$\left. \begin{aligned} \text{var}(X_{ij}) &= p_1(1 - p_1) \\ \text{cov}(X_{i1}, X_{i2}) &= \alpha p_1(1 - p_1) \end{aligned} \right\} \tag{4.4.5}$$

and

$$\text{cov}(X_{i_1 j_1}, X_{i_2 j_2}) = \psi p_1(1 - p_1) \quad i_1 \neq i_2,$$

where ψ is the probability of identity by descent from the foundation stock of genes in different individuals, the kinship coefficient (section 2.2). Thus

$$\text{var}\,(\bar{X}_{..}) = \left(\frac{1}{2N}\right)\{1 + \alpha + 2(N-1)\psi\}p_1(1-p_1)$$

$$= \psi_{\text{W}}p_1(1-p_1), \qquad (4.4.6)$$

where ψ_{W} is the within-group coancestry (Cockerham, 1967).

4.4(ii) *A hierarchic model*

Proceeding now to the next level of the hierarchy, consider a collection of groups, and define X_{kij} as in (4.4.4) for the jth gene ($j = 1, 2$) of the ith individual ($i = 1, \ldots, n_k$) of the kth group ($k = 1, \ldots, K$). Let $N = \sum_{k=1}^{K} n_k$ be the total number of individuals. The variability may be analysed by means of the linear model

$$X_{kij} = p_1 + \gamma_k + \beta_{ki} + \varepsilon_{kij}, \qquad (4.4.7)$$

where p_1 is the overall expected allele frequency, and effects γ_k, β_{ki} and ε_{kij} are random and uncorrelated with variances σ_{B}^2, σ_{BW}^2 and σ_{WW}^2 respectively, these being the variances between groups, between individuals within groups, and between genes within individuals within groups. Then $E(X_{kij}) = p_1$ and, for uncorrelated groups,

$$E(X_{kij}X_{k'i'j'}) - p_1^2$$

$$= \begin{cases} \sigma^2 = \sigma_{\text{B}}^2 + \sigma_{\text{BW}}^2 + \sigma_{\text{WW}}^2 & \text{if } k = k', i = i', j = j' \\ \sigma_{\text{B}}^2 + \sigma_{\text{BW}}^2 & \text{if } k = k', i = i', j \neq j' \\ \sigma_{\text{B}}^2 & \text{if } k = k', i \neq i' \\ 0 & \text{if } k \neq k' \end{cases} \qquad (4.4.8)$$

Further

$$\sigma^2 = p_1(1-p_1) \quad \text{for the variance of genes,}$$

$$\sigma_{\text{B}}^2 + \sigma_{\text{BW}}^2 = \alpha p_1(1-p_1) \quad \text{for genes within individuals,}$$

and

$$\sigma_{\text{WW}}^2 = \psi p_1(1-p_1) \quad \text{for genes within groups.}$$

Thus we have a partition of the total variance

$$\left. \begin{aligned} \sigma^2 &= \sigma_{\text{B}}^2 + \sigma_{\text{BW}}^2 + \sigma_{\text{WW}}^2 \\ (1-\alpha)p_1(1-p_1) &= \sigma_{\text{WW}}^2 \\ (\alpha - \psi)p_1(1-p_1) &= \sigma_{\text{BW}}^2 \\ \psi p_1(1-p_1) &= \sigma_{\text{B}}^2. \end{aligned} \right\} \qquad (4.4.9)$$

Correlations may be viewed over groups, within groups or within individuals, (Cockerham, 1973). Relative to the total population we have correlations between individuals within a group:

$$\psi = \sigma_B^2/(\sigma_B^2 + \sigma_{BW}^2 + \sigma_{WW}^2)$$

and between genes within individuals;

$$\alpha = (\sigma_B^2 + \sigma_{BW}^2)/(\sigma_B^2 + \sigma_{BW}^2 + \sigma_{WW}^2).$$

But we have also a correlation between genes within individuals relative only to their group

$$\alpha^* = \frac{\sigma_{BW}^2}{\sigma_{BW}^2 + \sigma_{WW}^2}, \quad \gamma_k \text{ now being a fixed effect,}$$

$$= \frac{\alpha - \psi}{1 - \psi}$$

or

$$(1 - \alpha) = (1 - \alpha^*)(1 - \psi), \tag{4.4.10}$$

which is analogous to the well-known formula of Wright (1951) for hierarchic structure:

$$(1 - F_{IT}) = (1 - F_{IS})(1 - F_{ST}), \tag{4.4.11}$$

where, in the notation of Wright (1951), I denotes individuals, S denotes subpopulations and T the total population.

Equation (4.4.8) also enables us to consider the heterozygosity amongst subpopulations. For this we require

$$\text{var}(\bar{X}_{k..}) = E(\bar{X}_{k..} - p_1)^2 \quad \text{where } \bar{X}_k = \frac{1}{2n_k} \sum_{i,j} X_{kij}$$

$$\left.\begin{aligned}
&= \frac{1}{2N}\sigma_{WW}^2 + \frac{1}{N}\sigma_{BW}^2 + \sigma_B^2 \\
&= \left\{\frac{(1-\alpha)}{2N} + \frac{(\alpha - \psi)}{N} + \psi\right\} p_1(1-p_1) \\
&= \psi_W p_1(1-p_1).
\end{aligned}\right\} \tag{4.4.12}$$

from (4.4.6) Hence the *allelic diversity* or heterozygosity under Hardy–Weinberg equilibrium is

$$H = E(2\bar{X}_{k..}(1 - \bar{X}_{k..})) = 2p_1(1-p_1)(1-\psi_W)$$

$$= 2\{p_1(1-p_1) - \text{var}(\bar{X}_{k..})\}$$

or

$$\text{var}(\bar{X}_{k..}) = p_1(1-p_1) - \tfrac{1}{2}H. \tag{4.4.13}$$

Equation (4.4.13) is a restatement of Wahlund's principle (Wahlund, 1928), var $(\bar{X}_{k.})$ being the variance in allele frequency amongst random groups. Note, however, that the heterozygosity H is not the frequency of heterozygotes expected in the population, but the frequency amongst a sample of N offspring under random union of gametes. Within the current population the expected frequency of heterozygotes is

$$H^* = E(2X_{kij}(1 - X_{kij'})) \quad j \neq j'$$
$$= 2p_1(1 - p_1)(1 - \alpha) = 2\sigma^2_{WW}.$$

Under random union of gametes $\alpha = \psi$ and

$$1 - \psi_W = \frac{(2N - 1)}{2N}(1 - \alpha)$$

or

$$H^* = 2NH/(2N - 1).$$

However, for a dioecious organism in which selfing is not possible, random mating is not equivalent to random union of gametes, and H bears no simple relation to H^*. Rewriting (4.4.12) as $\psi_W = $ var $(\bar{X}_{k.})/p_1(1 - p_1)$ we see that this within-group coancestry, ψ_W, is the ratio of actual variance in allele frequencies amongst groups to its limiting value, and hence always positive (see also Sections 4.5(ii) and 4.5(iv)).

4.4(iii) *Animal breeding systems*

In presenting the variance components via a linear model, variances (4.4.8) are necessarily positive. However, even under genetic drift and structural system alone, some intraclass correlations may be negative. If mates are less related than the average within the group, $\psi > \alpha$ and 'σ^2_{BW}' < 0. For example, the three statistics α, ψ and α^* (or in notation of Wright, F_{IT}, F_{ST} and F_{IS}) may be used to analyse the inbreeding developed in a subpopulation, which may be a breed or a line of a mating system. The total *group* of populations is the hypothetical infinite group that could arise from the founders under replications of the population process. Correlations α and ψ relative to this group are thus simply probabilities of gene identity by descent from the founders and are necessarily non-negative. However, at any stage in the development of the breed at which mates are less related than random individuals, $\alpha^* < 0$. In animal breeding there is no reason to suppose α^* will remain of the same sign throughout the process, but in some mating systems it does so. For simplicity we shall consider only sib-mating; other systems are considered by Wright (1965).

In sib-mating there are only two distinct correlations between the allelic types of two genes. There are correlations between the two genes within an individual at generation t, $\alpha^{(t)}$, and between two in different individuals, $\psi^{(t)}$. Further, since the genes in different individuals at generation t form the pairs within individuals at $(t+1)$,

$$\psi^{(t)} = \alpha^{(t+1)}. \tag{4.4.14}$$

From (4.2.5) we have also, noting (4.2.4),

$$\alpha^{(t)} = \tfrac{1}{2}\alpha^{(t-1)} + \tfrac{1}{4}\alpha^{(t-2)} + \tfrac{1}{4} \cdot 1,$$

and hence

$$(1 - \alpha^{(t)}) = \tfrac{1}{2}(1 - \alpha^{(t-1)}) + \tfrac{1}{4}(1 - \alpha^{(t-2)})$$

and

$$(1 - \alpha^{(t)}) = A\{\tfrac{1}{4}(1 + \sqrt{5})\}^t + B\{\tfrac{1}{4}(1 - \sqrt{5})\}^t, \tag{4.4.15}$$

where

$$A = \{1 - \tfrac{1}{4}(1 - \sqrt{5})\}/\{\tfrac{1}{4}(1 + \sqrt{5}) - \tfrac{1}{4}(1 - \sqrt{5})\}$$

and

$$B = (1 - A).$$

Since $\alpha^{(t)}$ is increasing, $\psi^{(t)} > \alpha^{(t)}$ and $\alpha^{*(t)} < 0$. In fact

$$\alpha^{*(t)} = 1 - 4\left\{ \frac{A(1 + \sqrt{5})^t + B(1 - \sqrt{5})^t}{A(1 + \sqrt{5})^{t+1} + B(1 - \sqrt{5})^{t+1}} \right\}$$

$$\rightarrow 1 - \frac{4}{(1 + \sqrt{5})} = \frac{\sqrt{5} - 3}{\sqrt{5} + 1} = -0.236. \tag{4.4.16}$$

Clearly in the case of sib-mating, genes chosen from mates have lower probability of identity than random genes at a given generation, for all pairs of genes are either in sibs (mates) or within an individual.

In any system of maximum avoidance (Section 4.2(iii)) we must again have $\alpha^* < 0$, since mates are, by definition, the least related pairs of individuals within the population. In cases of circular half-sib-mating or first-cousin mating, mates may be more or less related than the average, depending on the size of the circle. In fact only for a circle size four are half-sib mates less-than-averagely related, whereas for first-cousins this remains the case in a line of size eight, but not 16 (Wright, 1965). We here view the regular mating systems as a first approach to the more general problem of partial isolation of natural subpopulations. Although the relationship (4.4.10) between correlations relative to lines (α^*) and relative to a hypothetical total (α) is illuminating, we have only these two levels of hierarchy.

4.4(iv) *A general hierarchy*

In general, further levels of structure may be introduced into a population. As in (4.4.3) and (4.4.4) we define

$$X_{i_K,i_{K-1},\ldots,i_1,i_0} = \begin{cases} 1 & \text{if designated gene is } A_1 \\ 0 & \text{otherwise,} \end{cases}$$

where subscript $i_0 = 1, 2$ designates the two genes within an individual, and $i_1, \ldots, i_k$ designate components within K increasing levels of hierarchic structure. Analogously to (4.4.7), we may consider the linear model

$$X_{i_K,\ldots,i_1,j} = p_1 + \gamma_{i_K} + \beta_{i_K,i_{K-1}} + \cdots + \varepsilon_{i_K,\ldots,i_1,i_0},$$

all effects again being uncorrelated. Now (4.4.8) and (4.4.9) become

$$\sum_{l=j}^{K} \sigma_l^2 = \psi_{j-1} p_1(1-p_1), \quad (\psi_0 = \alpha) \tag{4.4.17}$$

where σ_l^2 is the variance of the lth-level effects, ψ_j the correlation at the jth level relative to the total, and α that between genes within individuals, or

$$\left.\begin{aligned} \sigma_j^2 &= (\psi_{j-1} - \psi_j)p_1(1-p_1) \quad 2 \le j \le (K-1) \\ \sigma_1^2 &= (\alpha - \psi_1)p_1(1-p_1) \\ \sigma_{\text{WW}}^2 &= (1-\alpha)p_1(1-p_1) \quad [\sigma_{\text{WW}}^2 = \text{var}(\varepsilon)] \\ \sigma_K^2 &= \psi_{K-1} p_1(1-p_1). \end{aligned}\right\} \tag{4.4.18}$$

Equations (4.4.6) and (4.4.12) now become

$$\text{var}(\bar{X}\ldots) = \left\{\frac{(1-\alpha)}{2N} + \frac{(\alpha-\psi_1)}{R_1} + \frac{(\psi_1-\psi_2)}{R_2} + \cdots + \frac{\psi_{K-1}}{R_K}\right\}p_1(1-p_1),$$

where $R_1, \ldots, R_K$ depend only on the numbers of genes in each subdivision at each level of the hierarchy (Cockerham, 1969).

More important is the extension of (4.4.10). Rather than correlations relative to the total, consider correlations relative only to the jth level of the hierarchy. Noting that

$$\sigma_{\text{ww}}^2 + \sum_1^i \sigma_i^2 = (1 - \psi_j)p_1(1 - p_1)$$

we have correlation

$$\rho_{j-1|j} = \sigma_j^2 \Big/ \Big\{ \sigma_{\text{ww}}^2 + \sum_1^i \sigma_i^2 \Big\} = (\psi_{j-1} - \psi_j)/(1 - \psi_j)$$

or

$$(1 - \psi_{j-1}) = (1 - \rho_{j-1|j})(1 - \psi_j). \qquad (4.4.19)$$

Denoting the within-individual level as zero, we may compute all overall correlations in terms of one-step correlations of each level relative to the next. From (4.4.19) we see

$$\left.\begin{array}{l}(1 - \psi_r) = \displaystyle\prod_{j=r}^{K-1} (1 - \rho_{j|j+1}) \quad r = 1, \ldots, K-1 \\[4mm] (1 - \alpha) = \displaystyle\prod_{j=0}^{K-1} (1 - \rho_{j|j+1}).\end{array}\right\} \qquad (4.4.20)$$

4.4(v)   *Isolation by distance*

A hierarchic structure is a first approach to the problem of isolation by distance: the partial subdivision of a population due to the tendency to mate within a geographic locality. In the theory developed above there are no assumptions about the evolutionary causes of differentiation, although, where correlations are identified with probabilities of identity by descent computed on basis of pedigrees, absence of differential selection is implicit. There is also no restriction on the degree of isolation between subpopulations at different levels of the hierarchy. They may be completely isolated *islands*, or merely arbitrarily bounded areas of a continuously distributed population. In the study of isolation by distance in a continuum (Wright 1940, 1943) the important unit is the *neighbourhood*, n. This is the size of population from which parents of individuals may be assumed to be drawn at random. For simplicity we shall consider only the simplest case of random union of gametes. Other more complex systems of mating do not give qualitatively different results (Wright, 1946).

In the case of an area continuum, in which birthplaces of offspring are normally distributed relative to those of parents, parents of individuals in a localized neighbourhood of size rn are drawn at random from a neighbourhood of size $(r+1)n$ (Wright, 1943). Let ψ_r be the correlation between gametes drawn at random from subpopulation size rn, relative to the total of size Kn. Then

$$\psi_r^{(t)} = \frac{1}{rn}\{\tfrac{1}{2} + \tfrac{1}{2}\psi_1^{(t-1)}\} + \left(1 - \frac{1}{rn}\right)\psi_{r+1}^{(t-1)}, \tag{4.4.21}$$

since with probability $1/rn$ the gametes come from the same individual, in which case they are the same gene with probability $\tfrac{1}{2}$ and the two different genes with probability $\tfrac{1}{2}$. In the latter case they are random genes from population size n at the preceding generation. With probability $1 - 1/rn$ the two genes are in different individuals, and hence parental gametes are randomly chosen from population size $(r+1)n$ at the preceding generation.

The equilibrium solution to equations (4.4.21) may be readily obtained, the correlation of interest being the within neighbourhood correlation ψ_1:

$$\psi_1 = \sum_{r=1}^{K-1} \theta_r \Big/ \left(2n - \sum_{r=1}^{K-1} \theta_r\right), \tag{4.4.22}$$

where

$$\theta_r = \left\{\frac{(r-1)n - 1}{rn}\right\}\theta_{r-1}$$

(Wright, 1951). As K becomes large, that is relative to an infinite total, $\psi_1 \to 1$. If $K = 1$, $\psi_1 = 0$.

To measure the differentiation between subpopulations as the size of subpopulation considered becomes larger we require correlations within a subpopulation size Sn, say, relative to the total Kn. This can be obtained from correlations ψ_1 as follows. Let the correlation ψ_1 within neighbourhoods, relative to a total Kn, now be denoted $\psi_1^{(K)}$ and the required correlation $\rho_S^{(K)}$. Then as in (4.4.19)

$$(1 - \psi_1^{(K)}) = (1 - \rho_S^{(K)})(1 - \psi_1^{(S)})$$

or

$$\rho_S^{(K)} = \{\psi_1^{(K)} - \psi_1^{(S)}\}/(1 - \psi_1^{(S)}). \tag{4.4.23}$$

By plotting numerically values of $\rho_S^{(K)}$, computed from (4.4.22) and (4.4.23), against S (or $\log S$), as S ranges from 1 to K, we may investigate the decrease in differentiation as progressively larger arbitrarily defined regions are considered. When $S = 1$ we have the differentiation $\psi_1^{(K)}$ between natural neighbourhoods. When $S = K$ we have, of course, no differentiation. The case of a linear continuum may be considered similarly; in this case parents of individuals in a subpopulation size $\sqrt{r}n$ come from population size $\sqrt{(r+1)}n$ (Wright, 1951).

4.5 Variance of the population processes

In the previous section we have considered the partition of the variation observed in a population. This partition is independent of assumptions about the cause of variation, but if any *interpretation* of the components is required some population model must be superposed. Two types of inferences have been based on observed variability between populations. Morton *et al.* (1968, 1971) use equation (4.4.12) to estimate coancestry (gene-identity) between populations. Lewontin and Krakauer (1973) have used equilibria of the Wahlund variance (4.4.13) under models of isolation by distance (4.4.22) to test neutrality of alleles. Ewens (1973) has also provided a test of neutral allelic equilibrium based on within-population variability. For both these purposes (in the latter case, for the null hypothesis, at least) it is sufficient to consider variability between populations caused by (partial) isolation and random genetic drift alone.

Where inferences are to be based on observations it is insufficient to consider only expectations of a process under random genetic drift. Although the processes of gene-identity and gene-alikeness (allelic correlation) parallel each other in expectation, their variance is large, and Jacquard (1975) has commented on the need for confidence bounds on estimates of coancestry based on allelic correlation. A large variance need not preclude reliable estimation; the processes may have high correlation. That they are correlated is clear; in the absence of mutation, genes identical by descent are necessarily of the same allelic type. However, an analysis of the joint evolution of the processes (Thompson, 1976a) shows that the magnitude of the correlation is very small – certainly insufficient to guarantee any reliable estimate of coancestry.

4.5(i)   *The drift process in a random mating population*

We shall consider only the case of a single population, and the simplest model of a discrete-generation, random-union-of-gametes population of $2N$ genes (N individuals). Additional mating structure or partial subdivision will not provide qualitatively different conclusions. The evolution of measures of gene and allelic identity can both be studied via the process

$$K^{(t)} = \{k_{i,j}^{(t)}; \, 1 \le j \le r_i, \, 1 \le i \le s\},\qquad(4.5.1)$$

where initially there are r_i copies of allele A_i, $1 \le i \le s$, and $k_{i,j}^{(t)}$ is the number of replicates present at generation t of the jth initial gene of ith allelic type. Given $K^{(t)}$, the current structure of the population, the level of gene identity $F^{(t)}$ is the probability that two distinct genes randomly chosen from the current population are identical by descent from the $t = 0$ generation:

$$F^{(t)} = \sum_i \sum_j k_{i,j}^{(t)}(k_{i,j}^{(t)} - 1)/2N(2N - 1)$$

or

$$(1 - F^{(t)}) = \left\{1 - \frac{1}{2N}\right\}^{-1}(1 - G_2^{(t)}),\qquad(4.5.2)$$

where $G_2^{(t)} = \sum_i \sum_j (k_{i,j}^{(t)}/2N)^2$ is the second moment of *gene*-frequency.

Also heterozygosity can be measured as the probability that two randomly chosen distinct genes are of different allelic type:

$$H^{(t)} = \sum_i k_i^{(t)}(2N - k_i^{(t)})/2N(2N - 1)$$

$$= \left\{1 - \frac{1}{2N}\right\}^{-1}(1 - S_2^{(t)}),\qquad(4.5.3)$$

where $k_i^{(t)} = \sum_{j=1}^{r_i} k_{ij}^{(t)}$ is the current number of A_i alleles in the population, and $S_2^{(t)} = \sum_i (k_i^{(t)}/2N)^2$ is the second moment of the *allele* frequencies.

The second moments of processes $H^{(t)}$ and $F^{(t)}$, and their covariance, may now be investigated via the (up to fourth) moments of $k_{i,j}^{(t)}$ given by the multinomial sampling of the Wright model. The order of the mean changes in H and F over a single generation are of order N^{-1}, while in general the standard deviation is of order $N^{-1/2}$. Even in small populations the standard deviations of increments in the processes may be larger than the mean changes, while comparatively the situation becomes worse as N increases. Some numerical values are given by Thompson (1976a). There is one exception. When all genes have the same frequency var $(\delta F) =$ var $(F^{(t+1)}|K^{(t)})$ is of order $1/N^2$. This is the case at $t = 0$; $k_{ij}^{(0)} = 1$ for all i, j. Similarly, when all alleles have the same frequency, var (δH) is of order $1/N^2$. Thus in the initial generations var (δF) is relatively small, as also is var (δH) if initial allele frequencies are equal, but this situation deteriorates rapidly as t increases.

The covariance of H and F may also be approached via increments in the two processes conditional on $K^{(t)}$. From (3.2.4)

$$E(H^{(t+1)}|K^{(t)}) = \left(1 - \frac{1}{2N}\right)H^{(t)}$$

and

$$E(1 - F^{(t+1)}|K^{(t)}) = \left(1 - \frac{1}{2N}\right)(1 - F^{(t)})$$

and thus

$$E\{(H^{(t+1)} - E(H^{(t+1)}|K^{(t)}))(F^{(t+1)} - E(F^{(t+1)}|K^{(t)}))|K^{(t)}\}$$

$$= E(H^{(t+1)}F^{(t+1)}|K^{(t)}) - \left(1 - \frac{1}{2N}\right)H^{(t)}\left\{1 - \left(1 - \frac{1}{2N}\right)(1 - F^{(t)})\right\}.$$

$$(4.5.4)$$

This covariance is in general also of order N^{-1}, and the correlation can be over the whole range from -1 to $+1$. The correlation will normally be negative (H decreases as F increases) but, where H and $(1 - F)$ are in discord, the correlation in increments will be positive. For example, in the case of two alleles, where one consists of identical by descent replicates of one original gene, but the other allelic class contains only single replicates of distinct original genes, the correlation in increment is always positive and approaches $+1$ as $N \to \infty$ if the frequency of the first allele is greater than $\frac{1}{2}$. It is perhaps not surprising that there exist values of $K^{(t)}$ producing positive covariance in the increments of H and F. The relevant question is whether these values of $K^{(t)}$ will arise in practice. Thompson (1976a) found that some period of such discord between

H and $(1-F)$ arises during the evolution of the majority of small $(N \leq 200)$ populations.

From the variances and covariance of increments the cumulative second moments may be computed recursively using the formula

$$\text{var}\,(H^{(t)}) = \text{var}\,\{E(H^{(t)}|K^{(t-1)})\} + E\{\text{var}\,(H^{(t)}|K^{(t-1)})\}$$

$$= \left(1 - \frac{1}{2N}\right)^2 \text{var}\,(H^{(t-1)}) + E(\text{var}\,(\delta H^{(t)}|K^{(t-1)})). \quad (4.5.5)$$

The second term involves expectations of the fourth moments of allele frequencies at time $(t-1)$, and hence explicit formulae are cumbersome, but a numerical study at least may be made. Formulae similar to (4.5.5) hold also for

$$\left.\begin{array}{l} \text{var}\,(F^{(t)}) = \text{var}\,(1 - F^{(t)}) \\[2mm] \text{and} \\[2mm] \text{cov}\,(H^{(t)}, (1 - F^{(t)})). \end{array}\right\} \qquad (4.5.6)$$

Since H declines eventually to zero, and $H^{(0)}$ is assumed given, $\text{var}\,(H^{(t)})$ increases from zero and eventually declines again to zero. $\text{var}\,(F^{(t)})$ follows a similar pattern. The cumulative correlation between H and F is always negative (see below), but often very small. Thompson (1976a) gives numerical results.

4.5(ii) *Kinship and heterozygosity*

We have encountered several equations relating genotypic frequencies in infinite populations to probability of gene identity by descent, $\alpha^{(t)}$, relative to some fixed time $t = 0$. In a finite population these equations relate only to the expectation over the process of evolution:

$$\alpha^{(t)} = E(F^{(t)}|F^{(0)} = 0).$$

Thus (4.4.1) and (4.4.2) become

$$\left.\begin{array}{l} E(P_{ii}^{(t)}) = p_i^2 + \alpha^{(t)}p_i(1 - p_i) \\[2mm] \text{and} \\[2mm] E(P_{ij}^{(t)}) = 2(1 - \alpha^{(t)})p_i p_j \quad (i \neq j) \end{array}\right\}$$

or

$$E(H^{(t)}) = \sum_{i<j}\sum E(P_{ij}^{(t)})$$

$$= (1 - E(F^{(t)}|F^{(0)} = 0))H^{(0)} \qquad (4.5.7)$$

$\{p_i\}$ being the initial allele frequencies. Equation (4.2.4) becomes

$$E(\mathbf{G}^{(t)}) = (1 - E(F^{(t)}|F^{(0)} = 0))\mathbf{G}(\text{HW}) + E(F^{(t)}|F^{(0)} = 0)\mathbf{G}(I),$$

giving, in particular, the same equation (4.5.7) for the frequency of heterozygotes. In fact, more than (4.5.7) is often true: the equation

$$E(H^{(t)}|F^{(t)}, F^{(0)} = 0, H^{(0)}) = (1 - F^{(t)})H^{(t)} \qquad (4.5.8)$$

holds for those processes of evolution (and only those) in which all those $K^{(t)}$ obtainable from each other by permutations of their elements are equiprobable (Thompson, 1976a).

Where (4.5.8) holds we may consider further the cumulative covariance of $H^{(t)}$ and $F^{(t)}$.

$$E(H^{(t)}F^{(t)}) - E(H^{(t)})E(F^{(t)})$$
$$= H^{(0)}E((1 - F^{(t)})F^{(t)}) - H^{(0)}E(1 - F^{(t)})E(F^{(t)})$$
$$= -H^{(0)} \operatorname{var}(F^{(t)}). \qquad (4.5.9)$$

Thus the cumulative covariance of the processes H and F is always negative and follows the same pattern as the variance of F. Further

$$\rho(H, F) = \operatorname{cov}(H^{(t)}, F^{(t)})/\{\operatorname{var}(F^{(t)}) \operatorname{var}(H^{(t)})\}^{1/2}$$
$$= -H^{(0)}\{\operatorname{var}(F^{(t)})/\operatorname{var}(H^{(t)})\}^{1/2}.$$

. The net correlation between H and F thus increases, at least initially, but is often small, particularly in large populations where $\operatorname{var}(F^{(t)})$ remains small for some time, (see Section 4.5(i)). It also follows from (4.5.9) that the expected covariance between H and $(1 - F)$, at any point in time taken over the evolution process, is always positive. This is not inconsistent with periods of negative covariance in increments occurring at some stage in the evolution of many small populations.

Where (4.5.8) holds it is useful to consider also the process

$$Q^{(t)} = H^{(t)}/(1 - F^{(t)}) \qquad (4.5.10)$$

whose expectation, taken over all evolutions from $t = 0$, is constant:

$$E(Q^{(t)}) = E(E(Q^{(t)}|F^{(t)})) = E(H^{(0)}) = H^{(0)} = Q^{(0)}. \qquad (4.5.11)$$

To study the process we use the Taylor series expansion

$$\delta Q = \frac{\delta H}{(1 - F)} + \frac{H \, \delta F}{(1 - F)^2} + \frac{\delta H \, \delta F}{(1 - F)^2} + \frac{H \, \delta F^2}{(1 - F)^3} + \cdots$$

Hence, to order $1/N$ (recalling H and F are quadratic functions of

$\{k_{i,j}^{(t)}/N\}$), given the current H and F,

$$E(\delta Q) = \frac{E(\delta N \, \delta F)}{(1-F)^2} + \frac{HE(\delta F^2)}{(1-F)^3}$$

and

$$E(\delta Q^2) = (1-F)^{-4}\{(1-F)^2 E(\delta H^2) + 2H(1-F)E(\delta H \, \delta F) + H^2 E(\delta F^2)\}.$$

(4.5.12)

The increments in Q may thus be studied. It is of theoretical interest at least that $E(\delta Q)$ is of order $1/N$ except when $F = 0$. Thus, although over all possible evolutions the mean of $Q(t)$ is at all times equal to $Q^{(0)} = H^{(0)}$, in the course of any given evolution Q has non-zero drift. Equation (4.5.8) holds only for equiprobable permulations of the elements of $K^{(t)}$, and thus under our model this holds only for evolution relative to time 0 ($F^{(0)} = 0$, $k_{i,j}^{(0)} = 1$).

Of more practical importance is the variance of the process Q; although the within-evolution mean drift is non-zero it is very small. The estimation of coancestry from heterozygosity is, in effect, an assumption of constant Q. However, the standard deviation of Q is large, in fact similar to that of H. The numerical results of Thompson (1976a) relate to populations of size 100 to 500 and to periods of evolution up to $N/10$ (5 to 50 generations). For many problems of differentiation between and within groups of human populations this is the period of interest. For larger populations, results are qualitatively similar, at a given t/N. We have considered only a simple model of a single random-mating population, but the conclusions will hold in more general situations. The large variance of Q means that any estimate of ancestral relationship based on allelic correlations between populations will have wide confidence bounds. In practice, estimates of coancestry are not based on a single locus, but the results from several are averaged. Although reducing sampling variations, this creates a further statistical problem. Different loci have the same expected values of gene identity by descent between and within populations, but the actual $F^{(t)}$ estimated from the separate $H^{(t)}$ (or interpopulation equivalents) will differ. In this case we have therefore not only the component of uncertainty attributable to the large variance of Q (or of H at given F), but also the variance of F at different loci, about its expected value.

4.5(iii) *Several populations*

Nei and Chakravarti (1977) and Nei, Chakravarti and Tateno (1977) have adopted a similar approach to the problem of inferences based on population variation, but with reference to the problem of detecting selection. Consider a set of s subpopulations, the allele frequency of a certain allele being p_i in population i ($1 \le i \le s$). Let

$$\bar{p} = \frac{1}{s} \sum_{1}^{s} p_i$$

and

$$\sigma^2 = \frac{1}{s} \sum_{1}^{s} (p_i - \bar{p})^2,$$

and define

$$F_{ST} = \sigma^2 / \bar{p}(1 - \bar{p}) \qquad (4.5.13)$$

(cf. (4.4.11)). F_{ST} is thus the within-subpopulation correlation in allelic type relative to the total group (Wright, 1951). For an infinite set of populations σ^2 is the underlying variance, σ_0^2 say, in allele frequency, and $\bar{p}$ the underlying mean p_0. F_{ST} is thus the within-subpopulation coancestry, ψ_W (see (4.4.12)). For a finite subgroup of populations, we shall expect σ^2 to be an underestimate of the true infinite-group variance by a factor $(s-1)/s$. The denominator is also a biased estimate of the corresponding true infinite-group value:

$$\left. \begin{aligned} E(\bar{p}(1-\bar{p})) &= p_0(1-p_0) - \frac{1}{s}\sigma_0^2 \\ E(\sigma^2) &= \frac{s-1}{s}\sigma_0^2. \end{aligned} \right\} \qquad (4.5.14)$$

Lewontin and Krakauer (1973) suggest that a test of selection may be based on the coefficient of variation of F_{ST} amongst loci. Specifically they suggest

$$k = (s-1) \frac{\text{var}(F_{ST})}{E^2(F_{ST})}, \qquad (4.5.15)$$

should be approximately 2 under neutrality. This value is that given by a binomial underlying distribution of allele frequencies. If the distribution is flatter, or U-shaped, the allele frequencies between populations are more dispersed, but the *variance* of F_{ST} values between loci is reduced, and so also is k. On the other hand, a very heterogeneous set of F_{ST} values, and large k, can only be explained by the action of selection at some loci (Cavalli-Sforza, 1966).

To investigate Lewontin and Krakauer's proposal, Nei and Chakravarti (1977) have examined the variance of the statistic F_{ST} for a group of isolated populations. Although in (4.5.13) only a diallelic locus is considered, it is, in general, convenient to have a measure of variability amongst populations applicable to multiple alleles. Such a measure was introduced by Nei (1973) and may be computed as a weighted average of the F_{ST} values:

$$G_{ST} = \sum_{i=1}^{k} \bar{p}_j (1 - \bar{p}_j) F_{ST}^{(j)} \bigg/ \sum_j \bar{p}_j (1 - \bar{p}_j),$$

where $F_{ST}^{(j)}$ is the statistic (4.5.13) computed for the jth allele $(1 \le j \le k)$ and $\bar{p}_j$ the mean frequency of this allele over the group of populations, when $k = 2$, $G_{ST} = F_{ST}$. G_{ST} may also be written in the form

$$G_{ST} = 1 - \frac{H_S}{H_T}, \qquad (4.5.16)$$

where

$$H_S = \frac{1}{s} \sum_{i=1}^{s} \left(1 - \sum_{j=1}^{k} p_{ij}^2 \right)$$

is the average heterozygosity (allelic-diversity) within populations, and

$$H_T = \left(1 - \sum_{1}^{k} \bar{p}_j^2 \right)$$

is the overall allelic diversity, p_{ij} being the frequency of allele j $(1 \le j \le k)$ in population i $(1 \le i \le s)$.

The variance of the statistic G_{ST} may be investigated via a Taylor series expansion of (4.5.16) (cf. (4.5.12)). Expanding to include up to second moments of H_S and H_T, we have

$$E(G_{ST}) = 1 - \frac{E(H_S)}{E(H_T)} + \frac{\text{cov}(H_S, H_T)}{(E(H_T))^2} - \frac{E(H_S)\,\text{var}(H_T)}{(E(H_T))^3} \Bigg\}$$

and

$$\text{var}(G_{ST}) = \left\{ \frac{E(H_S)}{E(H_T)} \right\}^2 \left\{ \frac{\text{var}(H_S)}{(E(H_S))^2} + \frac{\text{var}(H_T)}{(E(H_T))^2} - \frac{2\,\text{cov}(H_S, H_T)}{E(H_S)E(H_T)} \right\},$$

$$(4.5.17)$$

(Nei and Chakravarti, 1977). The accuracy of this expansion is difficult to assess, the remainder terms being dependent not only on time and population size, but also on the initial allele frequencies. However (4.5.17) provides at least a first approximation. Let

$$
\left.
\begin{aligned}
J_S &= 1 - H_S = \frac{1}{s}\sum_i \sum_j p_{ij}^2 \\
J_T &= 1 - H_T = \sum_i p_{\cdot j}^2 \\
J_i &= \sum_j p_{ij}^2 \\
J_{ik} &= \sum_j p_{ij}p_{kj}.
\end{aligned}
\right\} \tag{4.5.18}
$$

The Js are thus measures of current allelic identity analogous to the gene identity F above. We have

$$
\left.
\begin{aligned}
E(J_S) &= E(J_i),\ E(J_T) = \{E(J_i) + (s-1)E(J_{ik})\}/s \\
\mathrm{var}\,(J_S) &= \mathrm{var}\,(J_i)/s \\
\mathrm{var}\,(J_T) &= \{\mathrm{var}\,(J_i) + 4(s-1)\,\mathrm{cov}\,(J_i, J_{ik}) + 2(s-1)\,\mathrm{var}\,(J_{ik}) \\
&\qquad + 4(s-1)(s-2)\,\mathrm{cov}\,(J_{ik}, J_{il})\}/s^3
\end{aligned}
\right\} \tag{4.5.19}
$$
and
$$
\mathrm{cov}\,(J_S, J_T) = \{\mathrm{var}\,(J_i) + 2(s-1)\,\mathrm{cov}(J_i, J_{ik})\}/s^2.
$$

As in the analysis of H and F above, these moments of J_i and J_{ik} are simple functions of the (up to fourth) moments of the within-population allele frequencies under random genetic drift, cf. (4.5.4). Since the populations are completely isolated, allele frequencies in different populations are independent. The required moments were given by Robertson (1952), and numerical values of the moments in (4.5.17) may be obtained.

Of particular interest is the case where the allele frequency is initially the same for all alleles, as in the case of analysis of $H(t)$ above. In this case

$$
\mathrm{cov}\,(J_i, J_{ik}) \equiv 0, \qquad \mathrm{cov}\,(J_{ik}, J_{il}) \equiv 0, \quad \forall t
$$
and
$$
\begin{aligned}
E(G_{ST}) &\simeq 1 - \frac{E(H_S)}{E(H_T)} \\
&= \frac{\left(1 - \dfrac{1}{s}\right)(1 - e^{-t/2N})}{1 - (1 - e^{-t/2N})/s},
\end{aligned} \tag{4.5.20}
$$

the remaining two terms of $E(G_{ST})$ being negligible. As

$$s \to \infty, \qquad E(G_{ST}) \to (1 - e^{-t/2N}), \qquad (4.5.21)$$

this being the allelic diversity in an infinite pool of populations (Wright, 1943). The relation between (4.5.20) and (4.5.21) is as expected on the basis of (4.5.14). As $s \to \infty$, var $(G_{ST}) \to 0$, but for small s ($s \leq 10$) we have a similar conclusion to that for $H^{(t)}$ and $Q^{(t)}$ above: for a range of initial allele frequencies, and times $t \leq 2N$, the standard deviation is of the same order of magnitude as the mean. For a pair of populations it is in fact larger than the mean. Thus any inference about the parameters of the process, for example time of divergence, based on the mean for infinite populations, (4.5.21), is likely to be misleading. Not only is the variance large, requiring a large number of independent loci to be used, but a correction for the mean (4.5.20) must also be made.

In addition Nei and Chakravarti compute the Lewontin and Krakauer statistic k (4.5.13) given by simulations of the process under their model of random genetic drift and isolated sub-populations. It is shown that, although initially less than 2, it may be substantially greater in later generations. For extreme initial allele frequencies ($p = 0.05$) this occurs even by $t \approx N/2$, and is due to the leptokurtic form of the gene frequency distribution. (Jacquard (1974) shows that the result of a leptokurtic distribution is to increase k, although his formula generally gives an overestimate (Nei and Chakravarti, 1977)).

4.5(iv) *Effect of migration*

In the case of isolated populations, G_{ST} (or F_{ST}) is the ratio of variability to its asymptotic value, when each of the sub-populations is fixed, and hence increases with time. When the population is only partially subdivided, only one allele can survive, and eventually all variation is lost. Thus, as in the case of $H(t)$ above, $F_{ST} \to 0$; expectations and variances first increase but eventually decline. This is not the case for an infinite collection of populations. In that case the variance of F_{ST} is of course zero, but the mean increases to an asymptotic limit. Consider, for example, the random-mating island model (Section 3.4), in which migration from the infinite population of mainland M occurs at rate β to every island subpopulation I of $2N$ genes at every generation. Then in the notation of Section 3.4, $p_S = 1/2N$, $H_{IM} = H_{MM} = 1$ and (3.4.1) becomes

$$(1 - H_{II}^{(t)}) = (1 - \beta)^2 \left\{ \frac{1}{2N} + \left(1 - \frac{1}{2N}\right)(1 - H_{II}^{(t-1)}) \right\}.$$

Since, relative to an infinite total pool of populations, F_{ST} is the probability of within-subpopulation indentity by descent (see (4.4.6), (4.4.12), (4.5.14)) it is here $(1 - H_{II})$ and

$$F_{ST}^{(t)} = (1 - \beta)^2 \left\{ \frac{1}{2N} + \left(1 - \frac{1}{2N}\right) F_{ST}^{(t-1)} \right\},$$

giving the asymptotic limit

$$F_{ST}^{(\infty)} = (1 - \beta)^2 / \{2N - (2N - 1)(1 - \beta)^2\} \qquad (4.5.22)$$

(cf. (3.4.3)) and approach to it given by

$$F_{ST}^{(t)} = F_{ST}^{(\infty)} - (F_{ST}^{(\infty)} - F_{ST}^{(0)}) \left\{ (1 - \beta)^2 \left(1 - \frac{1}{2N}\right)^t \right\}, \qquad (4.5.23)$$

(Wright, 1943). In the case of a finite number, s, of subpopulations we shall, as in (4.5.14), expect (4.5.23) to be an overestimate of expected observed normalized variability by approximately a factor $(s/s - 1)$.

Nei, Chakravarti and Tateno (1977) consider the case of a finite number of incompletely isolated populations, each subject to random genetic drift, via (within-population) random union of gametes and migration as in the island model. The evolution of the statistic F_{ST} (4.5.14) was considered. As in (4.5.12) and (4.5.17) approximate means and variances may be obtained by a Taylor series expansion of the ratio $\sigma^2/\bar{p}(1 - \bar{p})$. We again have symmetry between populations (see (4.5.19)), but we no longer have independence. Nevertheless a similar analysis may be performed, and theoretical results, together with Monte Carlo simulation, lead to similar conclusions. Standard deviations of F_{ST} are of the same order of magnitude as the means, and for $t \geq 2N$ are in fact larger. In the case of large β (0.5) and $t \leq 4N$, the mean value predicted by (4.5.23) agrees well with that observed for a small number of populations ($s = 5$) after correcting σ^2 by the factor $s/(s - 1)$. When m is small, or the initial allele frequency extreme, the finite-population case does not approximate the infinite-population situation even in the initial generations. In any analysis of data based on a finite set of populations some corrections should therefore be made.

Only a few studies of the variances of structural statistics of populations under evolutionary processes have been made, and these are preliminary. Models of random union of gametes within-populations are very much a first approach, as is the island model for populations that are not completely isolated. On the other hand, many analyses of data have been based on assumptions of equilibria, expectations and infinite-population theory. How far the inferences of these analyses can be justified remains uncertain, but these preliminary analyses of variance show that confidence bounds must be wide. More analyses of the variance of measures of populations structure under evolutionary pressure are required. Only then can the adequacy of estimates of evolutionary parameters, and underlying parameters of structure, be assessed.

5

Genealogical and genetic distance

5.1   Measures of distance

In previous chapters we have discussed coefficients of gene-identity, which parametrize joint probability distributions for genotypes of relatives and characterize population structure. Such coefficients may be used as a basis of measures of similarity or distance, between individuals and between populations. There are three basic aspects of any such measure. First, between what entities does it measure distance? Second, on what type of information is it based? Third, what is its purpose?

A measure of distance between populations will normally be based on genetic data: the allelic or genotypic frequencies within each population. Distance measures between individuals may be based either upon a genealogical relationship or upon genotypic data. If both types of information are available, we may base measures upon each, and may wish to consider to what extent they parallel each other, either in the context of a particular study or under some theoretical population model.

There are three possible aims in reducing data to a single pairwise measure of similarity or distance. First, as for the coefficient of kinship in Chapter 3 or for the correlation statistics of Chapter 4, we may have measures whose evolutionary behaviour under a variety of population models is of interest. Second, we may wish to obtain a simple representation of interrelationships via the distance measure, using multidimensional scaling for example. Finally, a distance measure based on observable genetic data may be used as a statistic in the estimation of underlying structure. Data on genotypes may be used to estimate genealogical relationships. Data on allele frequencies may be used to estimate the history of populations.

We consider first (Section 5.2) the estimation of genealogical relationship from genetic data. Although these estimates are not strictly *metrics*, their properties are fundamental to the relation between genealogy and genotypes, and hence to the development of distance measures based on one aspect having some relevance to the other. We shall then consider briefly (Section 5.3) the reconstruction of the complete genealogy of a population from genetic data. Here the class of potential estimates is open, and the question arises of an acceptable *representation* of the structure.

In Section 5.4 we consider a variety of genealogical and genetic distance measures proposed in the literature. Evolutionary properties of these measures have been previously studied; we consider their genetic relevance. In Section 5.5 we consider the problem of representation of genealogies via distance measures. Finally, in Section 5.6, we consider briefly the evolutionary, representational and estimation aspects of measures of distance between populations.

5.2   Estimation of genealogical relationship

For the analysis of phenotypic data, the appropriate measures of relationship are those coefficients of gene identity which parametrize the distribution of phenotypes amongst relatives (Section 2.6). Thompson (1975a) considers the estimation of these coefficients from genetic data.

Consider first the estimation of the three Cotterman coefficients (Section 2.3) on the basis of data for a pair of outbred relatives (Figures 5.2(a)–(e)). For phenotypes ϕ_1, ϕ_2 observed for individuals B_1, B_2 for a characteristic determined by a single autosomal locus (Figure 5.2(a)), we have from (2.3.1) and (2.5.1)

$$P(\phi_1, \phi_2|\mathbf{k}) = \sum_{i=0}^{2} k_i P_i(\phi_1, \phi_2), \qquad (5.2.1)$$

where $\mathbf{k} = (k_0, k_1, k_2)$ is the $\mathbf{k}$-vector for the genealogical relationship, and $P_i(\phi_1, \phi_2)$ is the probability of the ordered phenotype pair when the individuals have i genes in common. Thus $P_i(\phi_1, \phi_2)$ depends only on the characteristics (number of alleles, allele frequencies and patterns of dominance) of the locus. If now phenotypes $\phi_j^{(l)}$ ($j = 1, 2; 1 \leq l \leq r$) are observed at r unlinked autosomal

Figure 5.2(a). Contours of likelihood and expected log-likelihood in the space $K = \{(k_2, k_1, k_0); k_i \geq 0, k_2 + k_1 + k_0 = 1\}$. The true relationship is assumed to be sibs $[\boldsymbol{k} = (\frac{1}{4}, \frac{1}{2}, \frac{1}{4})]$, denoted by S. M denotes the maximum of the surface, and the arrow the direction of increasing contours. (a) Likelihood on the basis of one locus (M at vertex). (b) Likelihood on the basis of two loci (M on an edge of K). (c) General case; likelihood at three or more loci. (d) Expected log-likelihood; $M = S$. (e) Expected surface in the case of non-identifiability; $S \subset M$.

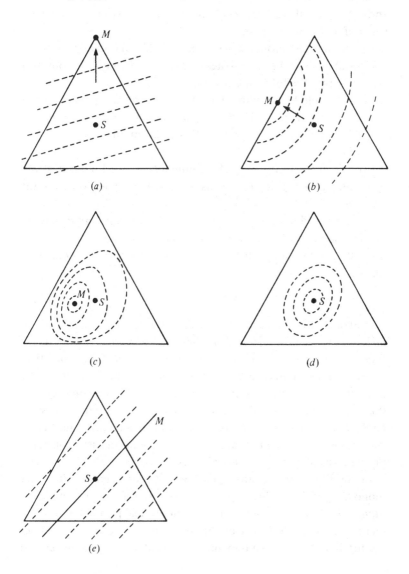

loci, we have likelihood function

$$L(\boldsymbol{k}) = \prod_{l=1}^{r} \left\{ \sum_{i=0}^{2} k_i P_i(\phi_1^{(l)}, \phi_2^{(l)}) \right\} \qquad (5.2.2)$$

for the vector of probabilities $\boldsymbol{k}$. This is a product of linear functions with non-negative coefficients. The log-likehood is thus concave, and the expected log-likelihood function has a maximum at the true value of $\boldsymbol{k}$ (Figure 5.2(d)).

In assessing the properties of maximum likelihood estimators of $\boldsymbol{k}$, consider first the uniqueness of this maximum. Thompson (1975a) gives the following two examples:

(i) a 3-allele locus, with allele A_1 dominant to A_2 dominant to A_3 dominant to A_1, and allele frequencies $p_1 = p_2 = p_3 = \frac{1}{3}$: for all phenotype pairs (5.2.1) is constant on lines of constant $k_0 - 2k_2$.

(ii) a 2-allele locus, with A_2 dominant to A_1: for all possible phenotype pairs (5.2.1) is constant on lines of constant $p_1 k_0 - k_2$.

Thus it is possible that, for certain loci, the parameters $\boldsymbol{k}$ may be unidentifiable, and certain pairs of relationship necessarily indistinguishable; in example (ii) we can never distinguish pairs $\boldsymbol{k}$ and $\boldsymbol{k}^*$ for which

$$p_1(k_0 - k_0^*) = (k_2 - k_2^*). \qquad (5.2.3)$$

The expected log-likelihood surface has a line of maxima, which includes the true value, but also other relationships (Figure 5.2(e)). Example (i) is a particular anomaly, but case (ii) is of some practical interest, since we may often wish to estimate relationships on the basis of diallelic loci exhibiting dominance. To do so we must ensure that they have widely differing frequencies, p_1, of the recessive allele. Different values of $\boldsymbol{k}$ are then (theoretically) distinguishable. (Relationships having the same $\boldsymbol{k}$ are clearly never distinguishable on the basis of genotypic data alone.)

In addition to identifiability we require information. A monomorphic locus clearly provides no information, and an *almost monomorphic* one almost none. The theoretical problem lies in the fact that different loci provide different contributions to the expected log-likelihood, since they have different allele frequencies and dominance patterns. From (5.2.2) we have

$$\log L(k) - \log L(k^*) = \sum_{l=1}^{r} Z_l(k, k^*), \qquad (5.2.4)$$

where

$$Z_l(k, k^*) = \log P(\phi_1^{(l)}, \phi_2^{(l)} | k) - \log P(\phi_1^{(l)}, \phi_2^{(l)} | k^*). \qquad (5.2.5)$$

We note $E(Z_l(k, k^*) | k) \geq 0$ for all k, k^* and all loci l. To adduce consistency in the estimation of k we require, not only that the number of loci becomes large, but that the number of *sufficiently informative* loci does so. For example, it is sufficient that the number of loci for which there exist positive ε, δ such that

$$P(Z_l(k, k^*) > \varepsilon | k) > \delta \qquad (5.2.5)$$

becomes large (ε and δ depending on k^*). Thompson (1975a) calls such loci (ε, δ)-*informative* with respect to the problem of distinguishing k and k^*.

In practice consistency is a rather artificial requirement; we cannot have data at unlimited numbers of unlinked loci. Nonetheless, the information in a locus, which can be measured by

$$E(Z_l(k, k^*) | k) \qquad (5.2.6)$$

and also by the set of (ε, δ) for which (5.2.5) holds, is important in any assessment of the power of genetic data to distinguish relationship k from k^*. Relationships well-spaced in the triangle K (Figure 5.2(a)–(e)), such as sibs, parent–offspring and unrelated pairs, can be readily distinguished from data on ten polymorphic blood-type systems. With more informative characteristics such as HL-A, we may also distinguish half-sibs, first-cousins and double-first-cousins. But this is about the limit of power of currently available data; some examples are given by Thompson (1975a).

The above theory extends directly to the estimation of arbitrary gene-identity-state probabilities. Equation (2.5.1) for arbitrary sets of individuals provides the analogue of (5.2.1), and for unlinked loci the likelihood is again the product of such linear terms. However the very large number of possible gene-identity-states creates both theoretical and practical problems. Those of non-identifiability are multiplied, and maximum likelihood estimates necessarily lie on boundaries of the total space. For estimation of k we see that a single locus provides always an estimate at the vertex of the triangle, and three loci are required for an internal maximum (Figures 5.2(*a*), (*b*) and (*c*)). In general we require *s* loci for *s* non-zero components of our estimate. Thompson (1975*a*) discusses these problems further, but here our purpose has been only to clarify further the interaction between genealogical relationship and genetic similarity.

5.3   Reconstruction of genealogies

Edwards (1967) first considered the reconstruction of a genealogy on the basis of genetic data. In theory, estimation of a complete genealogy is simply the inference of a joint genealogical relationship between individuals. We could consider all alternatives and compute a likelihood for each. However, specification of all genealogies between an ordered set of, say, 100 individuals is impractical. Instead, we must proceed sequentially, building up the genealogy from smaller units. A methodology applicable to the case of genotypic, rather than simply phenotypic, data has been given by Thompson (1976*d*).

Owing to symmetries in the likelihood for pairwise relationships, we can never infer which of a pair is uncle and which nephew (nor, in this particular instance, whether the pair are not half-sibs or grandparent–grandchild). Although in theory the problem is reduced by having further joint relatives, in practice it is necessary to have age data, or some other similar ordering of the individuals. We start construction from a set of unrelated individuals, and sequentially impose relationships between them, on the basis of the increase in log-likelihood provided. Where genotypes are available, offspring are conditionally independent given the parental geno-types, and the complete likelihood of a genealogy composed of two-parent nuclear families is the product of terms for the separate families. Denoting a proposed genealogy by $\mathscr{R}$ and the base-point of unrelated individuals by $\mathscr{U}$, we have relative log-likelihood

$$\log L(\mathscr{R}) - \log L(\mathscr{U})$$

$$= \sum_{j=1}^{m} \sum_{l=1}^{n_j} \log \{P(C_{jl}|M_j, F_j)/P(C_{jl})\}, \tag{5.3.1}$$

where $\mathscr{R}$ has m families, the jth having mother M_j, father F_j, and n_j offspring $C_{jl}(1 \le l \le n_j)$, and the individual label being used in this equation as a shorthand for the genotypic observations upon him.

Thus we wish to assign individual C to the pair (M, F) maximizing

$$\log \{P(C|M, F)/P(C)\}, \tag{5.3.2}$$

but simple application of this criterion is unlikely to lead to an accurate estimate of a human genealogy. Parents normally arise as couples having several offspring, and there are many restrictions upon marriage (Section 1.2). We therefore construct the genealogy by accepting whole nuclear families, maximizing

$$\sum_{l=1}^{n} \log \{P(C_l|M, F)/P(C_l)\}, \tag{5.3.3}$$

and incorporating any marriage restrictions sequentially into the set of potential marriages available for acceptance. Some of the offspring C_l may be later transferred to another accepted sibship, in which their contribution (5.3.2) is larger, but only contributions of individuals not already assigned to sibships are included in computation of criterion (5.3.3) for acceptance of a sibship.

Thompson (1976d) gives further details of the algorithm, which has been used to reconstruct genealogies accurately of up to 100 individuals on the basis of ten blood-group loci. Despite this success there is a major problem in that the algorithm determines most likely individuals to fulfill certain relationships, and not most likely relationships between specified individuals. It can in fact be shown that individuals M and F who provide maximum expected value of (5.3.2) are often not the true parents but the sibs of C (Thompson, 1976c). Thus, in a sense, the algorithm is inconsistent, and in practice relies heavily on age data, marriage rules and the determination of large sibships rather than simple triplets (C, M, F). These problems are magnified when only phenotypic data are available. Sibship phenotypes are not independent conditional on parental types, the overall log-likelihood is no longer the simple sum (5.3.1), and sibships separately providing high values of (5.3.3) may be jointly genetically incompatible. A sequential procedure may again be adopted, modifying log-likelihood contributions as construction proceeds, but problems arise in large genealogies.

If our aim is accurately to assign individuals to parents, the above algorithm has a high success rate. If it is to obtain an overall representation of the genealogical structure it is less satisfactory, for there is a tendency to determine only large sibships and to assign additional lone offspring to these. An accurately reconstructed genealogy is, by definition, the best representation of genealogical structure on the basis of genetic data. But, unless the algorithm provides a correct representation of the overall form, we must consider other general representations. Sibship size distribution is perhaps one of the most important criteria of overall form.

5.4   Measures of genealogical distance

We require a measure of distance, d, between members B_i of population $\mathscr{P}$, in order to represent the genealogical structure. Often d is required to have the basic properties of a metric

(a) $d(B_i, B_j) \geq 0$

(b) $d(B_i, B_j) = 0$

     if and only if $B_i = B_j$ for all $B_i, B_j, B_k \in \mathscr{P}$

(c) $d(B_i, B_j) = d(B_j, B_i)$

(d) $d(B_i, B_j) + d(B_j, B_k) \geq d(B_i, B_k)$

$$(5.4.1)$$

5.4.(i)   *Measures based on gene-identity*

Although the kinship coefficient (Section 2.2) is an intuitive simple measure of relationship, no function of $\psi(B_1, B_2)$ alone can satisfy (5.4.1), since $\psi(B, B)$ depends on $\psi(M, F)$, where M and F are the parents of B (see (2.2.2)). However, for any autosomal locus, the probability that two individuals have different genotypes does satisfy (5.4.1) (Jacquard, 1974), provided identical twins are allowed as an exception to (b). This measure is dependent upon the allele frequencies at the chosen locus. If we assume a highly polymorphic locus ($\sum_1^r p_i^2 \approx 0$) then we have

$$d_1(B_1, B_2) \approx 1 - \pi_1(B_1, B_2) - \pi_7(B_1, B_2), \qquad (5.4.2)$$

where $\pi_i(i = 1, \ldots, 9)$ are the probabilities of the nine identity states of Section 2.3, states 1 and 7 being those in which individuals necessarily have the same genotype (see Table 2.3 (a)).

A similar class of distance measures may be constructed as follows. Choose at random one of the allelic types represented amongst the four genes of the two individuals at the given autosomal locus. Jacquard (1974) shows that the probability that the allele is not present in *both* individuals satisfies (5.4.1). Considering again the limit for a highly polymorphic locus we have

$$d_2(B_1, B_2) = (\pi_2 + \pi_4 + \pi_6) + \tfrac{1}{2}(\pi_3 + \pi_5) + \tfrac{2}{3}\pi_8 + \pi_9. \quad (5.4.3)$$

The disadvantage of the above measures is that they are, in general, locus dependent. We must either accept the limiting forms d_1, d_2, or make arbitrary assumptions about allele frequencies.

The distance d_2 may, however, also be derived from genealogical rather than genotypic considerations. Rather than choosing at random a represented allele, we may choose an identity class of genes. Since in the polymorphic limit only genes identical by descent will be of the same allelic type, this gives again (5.4.3). Note that this may be rewritten

$$d_2(B_1, B_2) = 1 - \tfrac{1}{2}(\pi_3 + \pi_5) - \pi_7 - \tfrac{1}{3}\pi_8$$
$$= 1 - \{\psi + \tfrac{1}{2}\pi_7 + \tfrac{1}{12}\pi_8\} \quad (5.4.4)$$

(see (2.6.4) and (2.6.5)). This is the modification of ψ required to satisfy (5.4.1).

5.4(ii) *Measures based on genotype*

Where the genealogical relationship is unknown, the considerations leading to d_2 provide a measure based on genotype. We may now compute the probability that our randomly chosen representative allele is not present in both individuals over this

Table 5.4(a). *Pairwise metric based on genotype at an autosomal locus (see Section 5.4(ii))*

Genotypes		Type of randomly chosen allele				
B_1	B_2	A_i	A_j	A_k	A_l	Distance
A_iA_i	A_iA_i	1	—	—	—	0
A_iA_i	A_iA_j	$\tfrac{1}{2}$	$\tfrac{1}{2}$	—	—	$\tfrac{1}{2}$
A_iA_j	A_iA_j	$\tfrac{1}{2}$	$\tfrac{1}{2}$	—	—	0
A_iA_i	A_jA_j	$\tfrac{1}{2}$	$\tfrac{1}{2}$	—	—	1
A_iA_i	A_jA_k	$\tfrac{1}{3}$	$\tfrac{1}{3}$	$\tfrac{1}{3}$	—	1
A_iA_j	A_iA_k	$\tfrac{1}{3}$	$\tfrac{1}{3}$	$\tfrac{1}{3}$	—	$\tfrac{2}{3}$
A_iA_j	A_kA_l	$\tfrac{1}{4}$	$\tfrac{1}{4}$	$\tfrac{1}{4}$	$\tfrac{1}{4}$	1

If choice of alleles is weighted by their representation, the ordering of distances is unaltered random choice of alleles instead of over the alternative gene-identity-states. The resulting measure is given in Table 5.4(*a*). The expectation of this distance measure for individuals in given genealogical relationship is the previous measure of Section 5.4(i). If genotypes at several loci are observed, distances may be combined. For example, combination by averaging will ensure an overall measure satisfying (5.4.1).

A distance measure is a summary statistic on which a representation of the population may be based (Section 5.5). A measure based upon gene-identity may be so used directly, while one based upon genotype may either also be used directly or used to estimate first a genealogical distance function. In either case, a genotypic measure such as that proposed here, which has a direct parallel to a genealogical measure, is useful for purposes of comparison. As in Sections 5.2 and 5.3, the basic question is of the relationship between underlying genealogy and observable genotypes.

5.4(iii) *Measures based on ancestral contributions*

Although the gene-identity coefficients are the basic elements of genetic relationship, there are other bases for distance measures. Mycielski and Ulam (1969) propose the cardinality of the symmetric difference of the sets of ancestors of a pair of individuals within a defined pedigree. Any measure based on probabilities or cardinalities of set symmetric differences satisfies (5.4.1). Alternatively, the elements of ancestral sets may be restricted to the founders of the pedigree (e.g. Thompson, 1980*c*), or, viewing the pedigree from the founders, the distance between these may be defined to be the cardinality of the symmetric difference of sets of descendants.

More important than the number of ancestors are their genetic contributions. For example, Roberts (1971) considers the changing contributions of founders to a population's gene pool. Mycielski and Ulam (1969) define vectors $v(B)$ with components $v_{B*}(B)$ for all individuals B and B^* in $\mathcal{P}$:

(a) $v_{B*}(B) = 0$ if B^* is not an ancestor of B

                and $B^* \neq B$ (5.4.5)

(b) $v_{B*}(B) = \delta_{B*,B} + \tfrac{1}{2}\{v_{B*}(M) + v_{B*}(F)\},$

where

$$\delta_{B^*,B} \begin{cases} =1 & \text{if } B^* = B \\ =0 & \text{otherwise,} \end{cases}$$

and M and F are parents of B. The component $v_{B^*}(B)$ is uniquely defined by (5.4.5), for all B and B^*, and is the genetic contribution of B^* to B.

There are many ways in which a distance measure can be based on this genetic contribution. Mycielski and Ulam (1969) propose

$$d^{(\mathscr{P})}(B_1, B_2) = \sum_{B^* \in \mathscr{P}} |v_{B^*}(B_1) - v_{B^*}(B_2)| \tag{5.4.6}$$

while Lloyd (1973) extends this to the class of measures

$$d_k^{(\mathscr{P})}(B_1, B_2) = \left\{ \sum_{B^* \in \mathscr{P}} |v_{B^*}(B_1) - v_{B^*}(B_2)|^k \right\}^{1/k} \tag{5.4.7}$$

(in the notation of Mycielski and Ulam). Mycielski and Ulam consider the changing distribution of $d^{(\mathscr{P})}$ under some simple random pairing models for $\mathscr{P}$. Kahane and Marr (1972) generalize the class of population models, and obtain bounds on the expected values of the distance measure under these processes.

The sum of components of $v(B)$ is unbounded, which creates difficulties in the genetic interpretation of the measure and its practical use. A more genetically meaningful measure is obtained by restricting the indexing set of individuals B^* to the founders $\mathscr{F}$. Distances $d^{(\mathscr{F})}$ and $d_k^{(\mathscr{F})}$ may be defined precisely as in (5.4.6) and (5.4.7). Since

$$\sum_{B^* \in \mathscr{F}} v_{B^*}(B) = 1 \quad \text{for all } B \in \mathscr{P}$$

we have

$$\left. \begin{aligned} d^{(\mathscr{F})}(B_1, B_2) &\leq 2 \\ \text{and also} \\ d_k^{(\mathscr{F})}(B_1, B_2) &\leq 2^{1/k}. \end{aligned} \right\} \tag{5.4.8}$$

We may again also consider $\{v_{B^*}(B); B \in \mathscr{P}\}$, the descendant individuals B now becoming the indexing set. The sum $\sum_{B \in \mathscr{P}} v_{B^*}(B)$ is the expected contribution of ancestor B^* to the total population past and present, including himself. Edwards (1979) has considered these contributions as parameters of pedigree structure, while Jacquard (1974) proposes that a measure be based on contributions $\sum_{B \in \mathscr{P}_t} v_{B*}(B)$, to a particular generation t. This in turn may be used to develop a distance measure between populations $\mathscr{P}_t$.

5.4(iv) *Measures incorporating allelic change*

Another way to retain finite norm for vectors $v(B)$ is to replace (5.4.5) either by

$(a')\quad v'_{B*}(B) = \delta_{B,B*}\quad$ for founder individuals B

$(b')\quad v'_{B*}(B) = \lambda \delta_{B,B*} + \frac{1}{2}(1-\lambda)\{v'_{B*}(M) + v'_{B*}(F)\}$

$\qquad (0 < \lambda < 1),$

giving

$$\sum_{B* \in \mathscr{P}} v'_{B*}(B) = 1,$$

$$(5.4.9)$$

or by

$(b'')\quad v''_{B*}(B) = \delta_{B,B*} + \frac{1}{2}(1-\lambda)\{v''_{B*}(M) + v''_{B*}(F)\},$

giving

$$\sum_{B* \in \mathscr{P}} v''_{B*}(B) = 1/\lambda.$$

$$(5.4.10)$$

The former suggestion was made by Mycielski and Ulam, who comment 'perhaps a reasonable genetic interpretation is possible', and the latter by Lloyd (1973). In fact, $v'_{B*}(B)$ is the probability that a gene selected at random from B is of the allelic type *originating* as a new mutation in $B*$, where mutations at rate λ occur to new allelic types and founders each carry two unique alleles. On the other hand, $v''_{B*}(B)$ is the probability that the gene chosen from descendant B is *represented* in ancestor $B*$ under the same model.

Lloyd (1973) considers the inner product

$$v(B_1) \cdot v(B_2) = \sum_{B* \in \mathscr{P}} v_{B*}(B_1)v_{B*}(B_2) \qquad (5.4.11)$$

and a measure of relationship based on the modification (5.4.10):

$$r(B_1, B_2) = \tfrac{1}{2}v''(B_1) \cdot v''(B_2), \qquad (5.4.12)$$

which is related to distance measure $d_2^{(\mathscr{P})}$ of (5.4.7) by

$$\{d_2^{(\mathscr{P})}(B_1, B_2)\}^2 = 2\{r(B_1, B_1) + r(B_2, B_2) - 2r(B_1, B_2)\}. \qquad (5.4.13)$$

For vectors $v'(B)$, the inner product (5.4.11) is the probability that genes chosen from B_1 and B_2 originate as mutations in the same individual. Provided ancestors B^* are not inbred, for vectors $v''(B)$, (5.4.12) is the probability of choosing like alleles from B_1 and B_2. In general, however, it seems difficult to give a useful genetic interpretation to (5.4.10), (5.4.12) and $d_2^{(\mathcal{P})}$

5.4(v) *Modifications of the kinship coefficient*

The coefficient of relationship of Mallows (1974) is a modification of that of Lloyd (1973). It satisfies

$$r^*(B_1, B_2) = (\tfrac{1}{2} - \lambda)\{r^*(M, B_2) + r^*(F, B_2)\}, \qquad (5.4.14)$$

where M and F are parents of B_1, and B_1 is not an ancestor of B_2, and

$$r^*(B, B) = 1 \qquad (5.4.15)$$

for all B. A distance measure is given by $d^* = 1 - r^*$. Equation (5.4.14) provides r^* with properties similar to the kinship coefficient ψ (see (2.3.4)), particularly in the case of no mutation ($\lambda = 0$). Equation (5.4.15) circumvents the nonmetric objection to ψ, but also destroys the standard genetic interpretation. Thus the previous modification (5.4.4) may be preferable.

Lloyd and Mallows (1973) derive a distance measure based on Mendelian considerations, which may also be viewed as a modification of ψ. Consider a hypothetical diallelic locus, and score the genotypes $h(A_1A_1) = 1$, $h(A_1A_2) = 0$, $h(A_2A_2) = -1$. Then define

$$d_{\mathrm{LM}}(B_1, B_2) = E(h(B_1) - h(B_2))^2, \qquad (5.4.16)$$

where $h(B)$ is a shorthand for the score of the genotype of B, and expectations are taken over the random assignment of alleles to founders (allele frequencies p_1, p_2) and a mutation process. Then

$$d_{\mathrm{LM}}(B_1, B_2) = 4p_1p_2\{\mathcal{R}(B_1, B_1) + \mathcal{R}(B_2, B_2) - 2\mathcal{R}(B_1, B_2)\}$$
$$(5.4.17)$$

(cf. (5.4.13)), where $\mathcal{R}(B_1, B_2)$ is the correlation in allelic identity of genes chosen from B_1 and B_2, and where

$$\mathcal{R}(B_1, B_2) = \psi(B_1, B_2)$$

if there is no mutation. Lloyd and Mallows (1973) show that d_{LM} is a metric (provided identical twins are excluded in the case $\lambda = 0$), and (5.4.17) is perhaps the most useful metric based on ψ. Lloyd and

Mallows investigate properties of d_{LM}, and hence ψ, under a variety of random-mating models. Karlin (1968b) has investigated properties of ψ, and hence d_{LM}, under non-random mating.

5.5 Representation of genealogies

Several of the measures considered in the previous section have some relationship to both genetic and genealogical information. However, they do not permit detailed reconstruction of a genealogy, or estimation of genealogical relationship; for this many parameters are required (Section 5.2). They may allow representation of a genealogy, but for this purpose simpler metrics suffice. Although representations of a pedigree are not necessarily graphical, the purpose of a distance measure may often be to achieve a graphical representation of the structure.

The simplest method is to apply some cluster analysis method (Sokal and Sneath, 1963) to the individuals, on the basis of a distance measure, but there are few cases where this can achieve any clarification of the structure. More useful are methods which assign individuals to some point in space on the basis of the distances between them. One such is multidimensional scaling (Kruskal and Wish, 1978). To achieve a simpler representation of a large genealogy, one may prefer to consider larger units than individuals. For example, Thompson (1974b) obtained a representation of the Tristan da Cunha genealogy on the basis of multidimensional scaling of the mean kinship coefficients between sibships. If the distance measure allows individuals to be positioned exactly at points in a (high-dimensional) Euclidean space, methods of then reducing the dimension, such as principal components (Seal, 1964), are also available. Such methods are particularly useful for representations on the basis of genetic data. Data for different characteristics may be combined by first constructing coordinates on the basis of the separate characters and then using all dimensions jointly in obtaining the representation.

In some cases we may wish for a detailed representation of relationships within the pedigree. We may then find the pedigree more easily represented via the genealogical graph (Section 1.2). Summary parameters of this graph also provide useful measures of structure; for example, the diameter (Edwards, 1979), cutset sizes and in particular the minimax cutset (Thompson, Cannings and Skolnick, 1978). This last is an important measure of genealogical complexity, relevant to computational problems (Section 6.5). Coordinates based on the genealogical graph may also be used to investigate genealogical structure. Edwards (1979) has assigned coordinates on the basis of minimal node-graph distance of individuals from specified ancestors. The multidimensional histogram of individuals with respect to these coordinates can reveal structural features such as tenuously connected subsets. The positioning of individuals according to these coordinates can facilitate a drawing of the genealogy in traditional form (Edwards, 1979).

In addition to graphical representation, we may wish for a quantitative summary of pedigree structure. Ancestral contributions, mean inbreeding and kinship coefficients provide one such summary (Roberts, 1971). Another may be provided by the joint survival probabilities of sets of original autosomal founder genes (Thompson, 1979*b*). Within a specified genealogy, knowledge of the survival of certain founder genes decreases the survival probability of others. There are negative correlations between the survivals of founder genes, induced by Mendelian segregation in common descendants. The extent and pattern of correlation is a measure of structure of the genealogy, with respect to descent from founders. Thompson (1979*b*) has investigated the structure of the Tristan da Cunha genealogy by fitting a log-linear model to the extinction probabilities of all subsets of the set of original ancestral genes at a single autosomal locus. By partitioning the total variation into its main effects and separate interactions, the correlations in survival of ancestral genes can be precisely quantified. Further details are given by Thompson (1979*b*); here we present it only as an example of the way in which statistics computed on a genealogy may be used to achieve a non-graphical representation. The particular aspect of interest in this case was the ancestral origin of genes. In other cases other aspects will be relevant, and appropriate statistics should be chosen.

5.6   Distances between populations

Whereas measures of distances between individuals within a population may be based on either genealogical or genetic information, only the latter will normally be available when considering the larger-scale problem of relationships between populations. These genetic data will usually be the allele frequencies observed in different populations, at a variety of polymorphic loci. For more widely evolutionarily separated populations, such as species, the data may consist of the differing amino-acid sequences of proteins.

The most fundamental measure of relationship is the coefficient of kinship, and it is therefore natural to consider correlations of identity of allelic type of genes chosen from the separate populations (Section 4.4) as an approximation to the unobservable correlation in identity by descent. However, correlations computed from current allele frequencies are necessarily relative to some current gene pool (for example, the *least related members* in the current group; Cockerham, 1973). Although these *F*-statistics (Section 4.4) provide a good representation of interrelationships between populations, they cannot be expected to approximate genealogical estimates of common ancestry. Nei (1973) discusses this problem. Where ancestral allele frequencies are known, or equivalently overall population sizes very large, estimates of coancestry may be based upon allelic correlations (Morton *et al.*, 1971), but the confidence bounds in such estimates must be very wide (Section 4.5).

There are two distance measures which have been developed on the basis of models for the evolutionary history of populations. That of Cavalli-Sforza and Edwards (1967) was developed for its distributional properties under the process of random genetic drift. Distances between populations are statistics whose joint probability distribution under the model may be computed, for any specified historical relationships between the populations. Conversely, therefore, an estimate of these historical relationships is obtainable. Felsenstein (1968, 1973), Edwards (1970) and Thompson (1974*b*, 1975*b*) discuss this estimation problem in detail, and obtain estimation procedures.

The same distance measure has been widely used to obtain representations of relationships between populations. The metric allows the populations to be embedded in a multidimensional Euclidean space, and one representation of interrelationship is provided by the minimum spanning network of the extremal population nodes (Edwards and Cavalli-Sforza, 1963). This method can provide a useful illustration of relationship; Neel and Ward (1970) obtain such a representation of distances between American Indian villages. In using the distance measure in such an algorithm, there is no reason to infer that the result approximates an estimate of evolutionary history. In fact, Thompson (1975*b*) shows that this, and similar, algorithms are unlikely to provide an approximation to a good estimate of the evolutionary tree, because they do not ensure comparability of divergences of populations inferred to take place over the same time intervals. To interpret a representation of current dissimilarities between populations as an estimate of historical relationship, we require not only that the distance measure be based on a model for the process of divergence, but also that the estimation procedure be based on the probability distribution for the observable data under that model. The model of Cavalli-Sforza and Edwards (1967) is appropriate to separately evolving populations within a species, where genetic drift of allele frequencies can be assumed the major determinant of divergence.

Nei (1972) also proposes a measure of distance on the basis of an evolutionary model. The measure is a simple decreasing function of the correlation in allelic identity of genes chosen from different subpopulations (Section 4.5(ii)). Under a model of gene substitutions at the molecular level, the distance measure has a simple distribution dependent on the time of divergence of populations. Times of divergence may therefore be estimated; an example is given by Nei and Roychoudhury (1972). The model of Nei is appropriate for problems of long-time divergence between widely separated populations or between species, where gene substitution is a major effect. That of Cavalli-Sforza and Edwards should be applied to smaller-scale problems, where there are only small amounts of divergence, attributable to genetic drift. The probability distribution of their distance measure only takes a simple form when times of divergence are small relative to population size. Thompson (1975*b*) considers the adequacy of the approximations involved.

These measures of distance between populations are based on different considerations from those between individuals. For species the relevant factors will differ again. Distance measures may then be based on differences in the amino-acid sequences of proteins. But again there are the two approaches. Fitch and Margoliash (1967) provide a heuristic representation of relationships, by fitting the spanning network requiring the minimum total number of substitutions. Felsenstein (1968) considers the probability distribution of substitutions, under a mutation model, and estimates the historical structure on the basis of the model.

In all cases, the measures of distance providing satisfactory heuristic representations may differ from the statistics required in an estimation procedure. In the case of relationships between individuals, there is no simple set of such statistics, because of the complex way in which genealogy and genotypes are related (Sections 5.2 and 5.3). Measures for representational purposes (Sections 5.4 and 5.5) have little relevance to the estimation problem. For populations and species, the historical models are cruder, and the sufficient statistics therefore simpler. These statistics can therefore be directly used in heuristic representations of relationship. But we have seen that this does not guarantee that the representation approximates an estimate of evolutionary history on the basis of the model.

6

Algorithms

6.1 Basic recursions

In earlier chapters we have demonstrated some of the uses of coefficients of identity, and how recurrence relationships for these can be exploited to study, for example, population structure. Our purpose in this chapter is to discuss ways in which we may, for genealogies of varying complexity, carry out the evaluation of these coefficients.

The algorithms we shall describe are often based on the two recurrences

$$\psi(B_1, B_2) = \tfrac{1}{2}\psi(F_1, B_2) + \tfrac{1}{2}\psi(M_1, B_2), \qquad (6.1.1)$$

(see (2.3.7)) which is valid provided that B_1 is neither an ancestor of B_2 nor identical to B_2, and

$$\psi(B, B) = \tfrac{1}{2} + \tfrac{1}{2}\psi(F, M) \qquad (6.1.2)$$

(see (2.2.2)). For any given pair of individuals with a specified genealogy we can apply the above expressions recursively, as will be illustrated below, finally obtaining a numerical value. We will need, as always, to assume that $\psi(B_1, B_2) = 0$ if B_1 and B_2 are unrelated members of the genealogy, so that our result will be valid only within the frame of reference specified by the genealogy. Also $\psi(B, B) = \tfrac{1}{2}$ for an original individual.

6.2 Inbreeding coefficients

The coefficient of identity of two individuals is equal to the inbreeding coefficient of an offspring, so the calculations involved are equivalent for the two cases. Consider, as an example, the pedigree shown in Figure 6.2(a). Suppose we wish to calculate $\alpha(1) = \psi(2, 3)$. We repeatedly apply (6.1.1) and (6.1.2), and use $\psi(B, B) = \tfrac{1}{2}$ for B original, and $\psi(B_1, B_2) = 0$ if B_1 and B_2 are unrelated, to obtain the following sequence of equations (as for

Figure 6.2(*a*). Pedigree, taken from Stevens (1975).

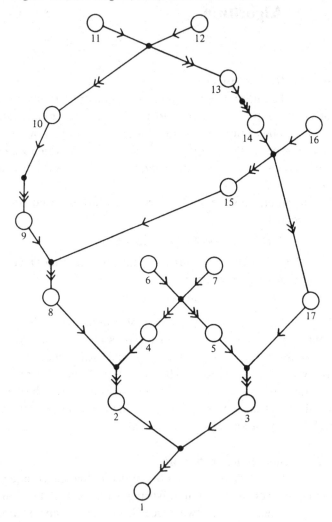

simpler examples in Section 2.3)

$$\alpha(1) = \psi(2, 3)$$
$$= \tfrac{1}{2}\{\psi(8, 3) + \psi(4, 3)\}$$
$$= \tfrac{1}{4}\{\underbrace{\psi(8, 5)}_{=0} + \psi(8, 17) + \psi(4, 5) + \underbrace{\psi(4, 17)}_{=0}\}$$
$$= \tfrac{1}{8}\{\psi(9, 17) + \psi(15, 17) + \psi(6, 5) + \psi(7, 5)\}$$
$$= \tfrac{1}{16}\{\psi(10, 17) + \psi(14, 17) + \psi(16, 17)$$
$$+ \underbrace{\psi(6, 6)}_{=\frac{1}{2}} + \underbrace{\psi(6, 7)}_{=0} + \underbrace{\psi(7, 6)}_{=0} + \underbrace{\psi(7, 7)}_{=\frac{1}{2}}\}$$
$$= \tfrac{1}{32}\{\psi(10, 14) + \underbrace{\psi(10, 16)}_{=0} + \underbrace{\psi(14, 14)}_{=\frac{1}{2}}$$
$$+ \underbrace{\psi(14, 16)}_{=0} + \underbrace{\psi(16, 14)}_{=0} + \underbrace{\psi(16, 16)}_{=\frac{1}{2}}\} + \tfrac{1}{16}$$
$$= \tfrac{1}{64}\{\psi(10, 13)\} + \tfrac{1}{32} + \tfrac{1}{16}$$
$$= \tfrac{1}{128}\{\psi(10, 11) + \psi(10, 12)\} + \tfrac{3}{32}$$
$$= \tfrac{1}{256}\{\underbrace{\psi(11, 11)}_{=\frac{1}{2}} + \underbrace{\psi(12, 11)}_{=0} + \underbrace{\psi(11, 12)}_{=\frac{1}{2}} + \underbrace{\psi(12, 12)}_{=0}\} + \tfrac{3}{32}$$
$$= \tfrac{25}{256} = 0.097\,656\,25. \tag{6.2.1}$$

It becomes clear on examining the equations (6.2.1) that we are essentially just tracing back the paths which the genes may have followed, until we come to a common point. All we have done is to identify the possible paths from individuals 11, 12, 16, 14, 6 and 7 which may result in identity between the genes of individual 1. We might therefore, alternatively, have identified in some way the possible paths from 1 to 1 through the pedigree. These paths are

1, 2, 4, 6, 5, 3, 1; 1, 2, 4, 7, 5, 3, 1; 1, 2, 8, 9, 10, 11, 13, 14, 17, 3, 1; 1, 2, 8, 9, 10, 12, 13, 14, 17, 3, 1; 1, 2, 8, 15, 14, 17, 3, 1

and

1, 2, 8, 15, 16, 17, 3, 1,

and correspond to the six terms $\psi(B, B)$ which arise in (6.2.1). The contribution of each of these paths is $(\tfrac{1}{2})^K$, where K is the number of individuals other than 1 in the path. The $(\tfrac{1}{2})^K$ represents the probability of a gene of the earliest member of the path being selected at each of the $(K+1)$ segregations involved (i.e. $(\frac{1}{2})^{K+1}$) multiplied by two as there are two genes in that individual. The discussion above illustrates the recurrence method and the method of paths, which is often a simpler and more efficient way of computing the result.

An algorithm for evaluating the inbreeding coefficient has been developed by Stevens (1975) using the enumeration of paths approach. Details of the algorithm, and a listing of a computer program for its implementation are given in his paper. One wishes to find all possible paths which start at one parent of 1, and finish at the other, subject to two restrictions: (a) that the sequence of individuals should proceed first from offspring to parent for a number of steps, and then from parent to offspring for the remainder (a sequence such as 1 2 8 9 10 12 13 14 15 16 17 3 1 is not permitted, the section 14 15 16 17 being a sequence which a gene cannot traverse); and (b) that no subsequence should be repeated in reverse order (a sequence such as 1 2 8 15 14 13 14 17 3 1 should be reduced to 1 2 8 15 14 17 3 1). An individual may occur twice in a sequence if he is inbred. These two rules guarantee that a gene can descend from the single common ancestor of the members of the sequence, and that no superfluous steps are included.

For the example under discussion we can proceed as follows (using the convention that one always proceeds upwards and to the left). Starting at 1 we proceed to 2, and thence to 8, to 9, to 10, to 11, and now start to come back down through 13, to 14 (not to 12 as this violates condition (a)), and so on. At any point where a sequence terminates, either by reaching 1 or by violating (a) or (b), one continues by generating a new sequence from the last point at which choice is still available.

Stevens also gives examples involving a more complex pedigree, from Chapman and Jacquard (1971), where finding the coefficient of inbreeding may involve a very large number of sequences (90 for a child of 1 and 3 for example).

6.3 General coefficients for a pair of individuals

As discussed in Chapter 2, the case of two individuals can involve one in the use of up to 15 coefficients of identity. One can (see for example Cockerham (1971)) write down analogues of

(6.1.1) and (6.1.2), but for the purposes of calculation, a generalization of the method of paths, as set out by Nadot and Vaysseix
(1973), is sufficient. Their method is similar in essence to that of
Stevens, generating all the possible sets of paths.

Consider the genealogy shown in Figure 6.3(*a*), and in particular
the two individuals 2 and 4. These two individuals possess four
genes 2^+, 2^-, 4^+ and 4^- being the paternal gene of 2, the maternal
gene of 2, the paternal gene of 4 and the maternal gene of 4,

Figure 6.3(*a*). Genealogy, after Chapman and Jacquard
(1971).

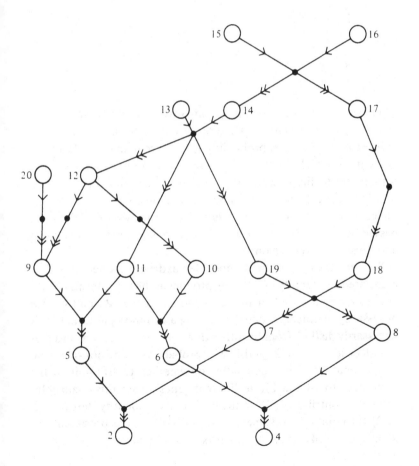

respectively. Following Nadot and Vaysseix we trace possible sequences (their *venues*) back from each of $2^+, 2^-, 4^+$ and 4^- to an original member (i.e. a founder) of the pedigree. For example $2^+, 5,$ 9, 21 is a possible sequence, which we will call a single-sequence.

In order to evaluate the probabilities of the various identity states for our two chosen individuals we need to combine the various single-sequences into joint sequences, specifying the routes of all the four genes. Such a joint sequence will be called a quadruple-sequence. Suppose, for example, that we have the four single-sequences $2^+, 5, 9, 20; 2^-, 7, 18, 17, 15; 4^+, 6, 10, 12, 14, 15$ and $4^-,$ 8, 18, 17, 15. These can be combined into a quadruple-sequence shown as

$$
\begin{array}{llll}
2^+ & 5 & 9 & 20 \\
2^- & 7 \\
4^- & 8
\end{array} \left.\begin{array}{l} \\ \\ \end{array}\right\} \quad 18 \quad 17 \quad 15
$$
$$
4^+ \quad 6 \quad 10 \quad 12 \quad 14 \quad 15
$$

where we have noted the common path involved in the descent of 2^- and 4^-. Now the gene 2^+ necessarily comes from 5 so that the descent path of 2^+ has probability $(\frac{1}{2})^2$ of occurring. Similarly the descent path of 4^+ has probability $(\frac{1}{2})^4$. The double-sequence corresponding to the descent of 2^- and 4^- has probability $(\frac{1}{2})^5$. We simply count the number of transitions involved in the passage of the gene through the sequence, noting that the event that the gene passed by 17 to 18 is that received from 15 is common to the paths and thus counted once only.

We have thus determined a possible quadruple-sequence, and an associated probability of $(\frac{1}{2})^{11}$, the product of the three terms above. We now observe that in this case the two genes 2^- and 4^- are necessarily identical by descent, having a common path, that 2^+ is necessarily distinct from the other three genes in question, and that 4^+ will be identical to 2^- and 4^- with probability $\frac{1}{2}$, and non-identical with probability $\frac{1}{2}$ (i.e. depending on whether 15 transmitted the same gene to 14 and 17, or different genes). Thus this quadruple-sequence contributes $(\frac{1}{2})^{12}$ to the probability of identity state 5 (1, 2, 2, 2) (in notation of Table 2.3(a) and $(\frac{1}{2})^{12}$ to the probability of identity state $14 \equiv (1, 2, 3, 2)$. An exhaustive enumeration of all the

quadruple-sequences, their probabilities and consequent contribution to the probabilities can proceed using the above ideas. Nadot
and Vaysseix have produced a computer program to implement the
algorithm. They actually achieve some reduction in the number of
sequences which need to be examined, by identifying common
ancestors, but the basic principle remains the above.

6.4   Multiple relationships

The extensions of the idea of coefficients of identity to more
than two individuals was discussed in Section 2.4, where it is also
shown how the specification of these can be simplified using the
notions of group theory. Algorithms have been developed by
Thompson (1974*a*) to exploit these ideas.

Consider the following example, discussed briefly by Thompson
(1974*a*). We have five individuals being three sibs, and their
maternal grandparents, as shown in Figure 6.4(*a*), where we have
also labelled the genes of the three founders of the genealogy. Our
first job is to specify the distinct orbits for the five individuals of

Figure 6.4(*a*). Example, after Thompson (1974*a*).

interest, viz. 1, 2, 5, 6 and 7. We first observe that we can, with no loss of generality, assign the genes 1 and 3 to the individual 4. The assignment of genes to 5, 6 and 7 can then be done in 64 ways, but as there is no necessity to distinguish genes 5 and 6 these reduce to 32, each with probability $\frac{1}{32}$, these being listed in Table 6.4(a). These 32 are the orbits required, and specify the 32 distinct identity states possible for the five individuals in question.

The use of these for the computation of a joint, or conditional, genotype (or phenotype) probability is relatively straightforward. Using again the example given by Thompson, suppose we wish to find the probability that the maternal grandparents were A and AB, given that the sibs were O, O and A, for the ABO blood group, when the gene-frequencies of A, B and O are p, q and r, respectively. It is first necessary to take a possible genotypic assignment for the individuals, say AO, AB, OO, OO, AO, and then to check which orbits this assignment is consistent with. For example, it is not consistent with the first orbit listed in Table 6.4(a), (1, 2, 3, 4, 1, 5, 1, 5, 1, 5) or with the last (1, 2, 3, 4, 3, 5, 1, 5, 1, 6). In this case there are

Table 6.4(a) *The genealogical relationship assigns probability $\frac{1}{32}$ to each of the following orbits (from Thompson, 1974a)*

(1,2,3,4,1,5,1,5,1,5)	(1,2,3,4,3,5,3,5,3,5)	identical sibs
(1,2,3,4,1,5,1,5,3,5)	(1,2,3,4,3,5,3,5,1,5)	
(1,2,3,4,1,5,3,5,1,5)	(1,2,3,4,3,5,1,5,3,5)	paternal gene in common
(1,2,3,4,3,5,1,5,1,5)	(1,2,3,4,1,5,3,5,3,5)	
(1,2,3,4,1,5,1,5,1,6)	(1,2,3,4,3,5,3,5,3,6)	
(1,2,3,4,1,5,1,6,1,5)	(1,2,3,4,3,5,3,6,3,5)	maternal gene in common
(1,2,3,4,1,5,1,6,1,6)	(1,2,3,4,3,5,3,6,3,6)	
(1,2,3,4,1,5,1,5,3,6)	(1,2,3,4,3,5,3,5,1,6)	
(1,2,3,4,1,5,3,6,1,5)	(1,2,3,4,3,5,1,6,3,5)	two sibs identical
(1,2,3,4,3,5,1,6,1,6)	(1,2,3,4,1,5,3,6,3,6)	
(1,2,3,4,1,5,1,6,3,5)	(1,2,3,4,3,5,3,6,1,5)	
(1,2,3,4,1,5,3,6,3,5)	(1,2,3,4,3,5,1,6,1,5)	
(1,2,3,4,1,5,1,6,3,6)	(1,2,3,4,3,5,3,6,1,6)	no two sibs identical
(1,2,3,4,1,5,3,5,1,6)	(1,2,3,4,3,5,1,5,3,6)	
(1,2,3,4,1,5,3,6,1,6)	(1,2,3,4,3,5,1,6,3,6)	
(1,2,3,4,1,5,3,5,3,6)	(1,2,3,4,3,5,1,5,1,6)	

five consistent assignments: (1, 2, 3, 4, 1, 5, 1, 5, 3, 6), (1, 2, 3, 4, 1, 5, 1, 5, 1, 6), (1, 2, 3, 4, 1, 5, 1, 5, 3, 5), (1, 2, 3, 4, 1, 5, 1, 6, 3, 5) and (1, 2, 3, 4, 1, 5, 1, 6, 3, 6), noting that we may rewrite AO as OA etc. The assignment (1, 2, 3, 4, 1, 5, 1, 5, 3, 6) has $1 \equiv O$, $2 \equiv A$, $3 \equiv A$, $4 \equiv B$, $5 \equiv O$, and $6 \equiv O$, i.e. a probability of $\frac{1}{32}p^2qr^3$. The joint probability can be accumulated by examining the other possible assignments and the required probability obtained by also finding the joint probability that the three sibs were O, O and A. Appropriate computer algorithms for carrying out these tasks have been written by Thompson (1974*a*), and used for up to six relatives. Obviously the computer time required increases rapidly with the number of relatives considered, and alternative methods are then needed, e.g. Section 6.5(iii).

6.5   Genealogies with phenotypic information

The methods described above are appropriate for genealogies when no information is available on the phenotypes of any of the individuals. On the other hand, we often have genealogies in which individuals are classified as normal, diseased etc., and may, for example, wish to calculate the risk that some prospective offspring will be affected. A number of workers (Hilden, 1970; Heuch and Li, 1972; Elston and Stewart, 1971; Cannings, Thompson and Skolnick, 1976, 1978, 1979, 1980; and Smith 1976) have addressed these problems, by essentially similar methods. The most general formulation to date allows the consideration of genealogies of (theoretically) arbitrary complexity, and a model of transmission involving correlations between relatives (through a linear model), assortative mating, age-specific and sex-specific penetrance, and various other features (Cannings, Thompson and Skolnick, 1980). The general treatment is by no means simple, and we shall give only two examples here which indicate the basic concepts, and the power of the method.

6.5(i)   *Pedigrees without loops*

Our first example concerns a simple pedigree (that is, one that has no loops in its graphical representation) shown in Figure 6.5(*a*). We suppose we are interested in some genetic character known to be inherited as a recessive. Phenotypic information is available on individuals 5 and 9, who are affected, and on 1, 2 and 3, who are unaffected. Suppose our aim is to find the probability that individual 6 (the *riskee*) has each of the three possible genotypes. We label the three genotypes 1, 2 and 3, the latter producing the affected phenotype. Our object is to calculate

$P\{$Individual 6 is affected$|$all available information$\}$.

Our method is a recursive one, relying on the fact that the removal of any non-peripheral individual will break the genealogy into pieces, which are interdependent only through the individual removed. In this particular case only individual 4 is non-peripheral; the removal of 4 breaking the genealogy into three distinct sets of individuals, (1 and 2), (3, 6 and 7) and (5, 8 and 9). Suppose that A is an event concerning the individuals 1 and 2, B one concerning 3, 6, and 7, and C one concerning 5, 8 and 9. Then A, B and C are mutually independent given the genotype of individual 4. We can exploit this fact to allow the successive condensation of the information in the genealogy in a way which will become clear through the steps illustrated below.

Figure 6.5(*a*). Pedigree without loops.

Key

⊗ unknown   ◯ unaffected   ◇ riskee   ⦸ affected

Consider first the removal of individuals 5, 8 and 9, and the condensation of all the information on three individuals into a function on individual 4. We shall denote by $E(i, j, \ldots, k)$ the event that the phenotypes of individuals $i, j, \ldots, k$ are as observed, and by $G_i(j)$ the event that the genotype of the ith individual is j. Then we can write

$$P(E(5, 8, 9) \mid G_4(i))$$

$$= \sum_j \sum_k \sum_l P(E(5) \mid G_5(j)) P(E(8) \mid G_8(k)) P(E(9) \mid G_9(l))$$

$$\times P(G_8(k) \mid G_4(i) \cap G_5(j)) P(G_9(l) \mid G_4(i) \cap G_5(j)) P(G_5(j)),$$

using straightforward algebraic manipulation, and exploiting the various independences of the events. We rewrite this as

$$P(E(5, 8, 9) \mid G_4(i))$$

$$= \sum_j \sum_k \sum_l \text{Pen}(5 \mid j) \text{Pen}(9 \mid l) \text{Trans}(k \mid i, j) \text{Trans}(l \mid i, j) \text{Init}(j),$$

where $\text{Pen}(K \mid j) = P(E(K) \mid G_K(j))$ for an individual K whose phenotype is observed i, i.e. it is the penetrance function, $\text{Trans}(l \mid i, j) = P(G_A(l) \mid G_B(i) \cap G_C(j))$ where individual A is the offspring of B and C, i.e. it is the Mendelian transition function for an autosomal locus, and $\text{Init}(j) = P$(an individual taken at random from the population has genotype j), which we take here to correspond to Hardy–Weinberg values. Each term of the above expression is known, and $P(E(5, 8, 9) \mid G_4(i))$ can be evaluated. In practice it will be somewhat easier to remove 8 and 9 one at a time; although in this case as 8 is unknown, and thus makes no contribution ($\text{Pen}(\mid)$ being one if the individual involved is of unknown type). Writing $R_4^*(i)$ for $P(E(5, 8, 9) \mid G_4(i))$, and $\boldsymbol{R}_4^* = (R_4^*(1), R_4^*(2), R_4^*(3))'$, we obtain

$$\boldsymbol{R}_4^* = \begin{bmatrix} 0 \\ \frac{1}{2}q^2 \\ q^2 \end{bmatrix},$$

where q is the frequency of the allele which in homozygous form has the *affected* type.

Out next step is to remove the individuals 1 and 2. This is achieved by finding $R_4^+(i)$ defined as

$$R_4^+(i) = P(E(1, 2) \cap G_4(i))$$

$$= \sum_j \sum_k P(E(1)|G_1(j))P(E(2)|G_2(k))P(G_4(i)$$

$$|G_1(j) \cap G_2(k))P(G_1(j))P(G_2(k))$$

$$= \sum_j \sum_k \text{Pen}(1|j)\,\text{Pen}(2|k)\,\text{Trans}(i|j, k)\,\text{Init}(j)\,\text{Init}(k).$$

We obtain for $\boldsymbol{R}_4^+ = (R_4^+(1), R_4^+(2), R_4^+(3))'$,

$$\boldsymbol{R}_4^+ = \begin{bmatrix} p^2(p+q)^2 \\ 2p^2q(p+q) \\ p^2q^2 \end{bmatrix},$$

the $(p+q)$s being retained, though equal to 1, to maintain the homogeneity of allele frequency exponent. We now combine the two functions $\boldsymbol{R}_4^*$ and $\boldsymbol{R}_4^+$ to obtain $\boldsymbol{R}_4$ say, using the equation $R_4(i) = R_4^+(i)R_4^*(i)$ obtained by the following argument (which relies on the conditional independence pointed out earlier).

$$R_4(i) = P(E(1, 2, 5, 8, 9) \cap G_4(i))$$

$$= P(E(5, 8, 9)|E(1, 2) \cap G_4(i))P(E(1, 2) \cap G_4(i))$$

$$= P(E(5, 8, 9)|G_4(i))P(E(1, 2) \cap G_4(i))$$

$$= R_4^+(i)R_4^*(i).$$

We have

$$\boldsymbol{R}_4 = \begin{bmatrix} 0 \\ p^2q^3(p+q) \\ p^2q^4 \end{bmatrix}.$$

Our final step is to compute $R_6(i) = P(G_6(i) \cap E(1, 2, 3, 4, 5, 9))$, individuals 7 and 8 having been removed as they are of unknown type.

We have

$$R_6(i) = P(E(1, 2) \cap E(5, 9) \cap E(4) \cap E(3) \cap G_6(i))$$

$$= \sum_j \sum_k P(E(1, 2) \cap E(5, 9) \cap E(4) \cap E(3) \cap G_6(i)$$

$$\cap G_3(j) \cap G_4(k))$$

$$= \sum_j \sum_k P(E(5, 9)|G_4(k))P(E(1, 2) \cap G_4(k))P(E(4)|G_4(k))$$

$$\times P(E(3)|G_3(j))P(G_6(i)|G_3(j) \cap G_4(k))P(G_3(j))$$

$$= \sum_j \sum_k R_4^* (k)R_4^+(k) \text{ Pen } (4|k) \text{ Pen } (3|j) \text{ Trans } (i|j, k) \text{ Init } (j).$$

We obtain

$$\boldsymbol{R}_6 = \begin{bmatrix} \frac{1}{2}p^3q^3(p+q)^2 \\ \frac{1}{2}(p+4q)(p+q)p^3q^3 \\ \frac{1}{2}p^3q^4(p+3q) \end{bmatrix}.$$

and so $P(E(1, 2, 3, 4, 5, 6, 9)) = \sum_{i=1}^3 R_6(i) = p^3q^3(p+2q)^2$, and the probabilities which we have as our objectives, are obtained by dividing the $R_6(i)$s by $p^3q^3(p+2q)^2$.

6.5(ii) *Pedigrees with loops*

We turn now to our second example. We again consider a disease which is inherited as an autosomal recessive, and consider the genealogy shown in Figure 6.5(*b*), which has a loop due to the marriage of uncle and niece (or aunt and nephew). We begin by

Figure 6.5(*b*). Pedigree with a loop.

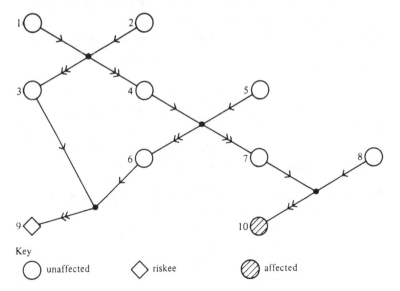

Key

◯ unaffected   ◇ riskee   ◪ affected

removing 8 and 10, and calculate

$$R_7(i) = P(E(8, 10)|G_7(i))$$
$$= \sum_j \sum_k P(E(8)|G_8(j))P(E(10)|G_{10}(k))$$
$$\times P(G_{10}(k)|G_7(i) \cap G_8(j))P(G_8(j)), \qquad (6.5.1)$$

where all summations are over the set of genotypes (here 1, 2 and 3), which may be evaluated since all of the probabilities on the right-hand side are specified by the model. The terms $P(E(8)|G_8(j))$ and $P(E(10)|G_{10}(k))$ are penetrances (here 0 or 1), $P(G_{10}(k)|G_7(i) \cap G_8(j))$ is the set of Mendelian segregation probabilities, and $P(G_8(j))$ is the set of population frequencies for the genotypes (we take p^2, $2pq$ and q^2 for 1, 2 and 3, p being the frequency of A). Thus we obtain

$$\boldsymbol{R}_7 = (R_7(1), R_7(2), R_7(3))' = (0, \tfrac{1}{2}pq, \phi),$$

ϕ indicating an expression later seen to be irrelevant.

Our next step, and we should emphasize that the steps are not rigidly defined, but to some extent optional, might be to 4 and 6, removing 7 and 5 at this step. Then the information on 8, 10, 7 and 5 will have been condensed onto 4 and 6. It will be necessary to use both 4 and 6, rather than only one, as the set chosen needs to split the genealogy into two disjoint pieces. The complete set of arcs between the section removed and that remaining, must have at one end one of the individuals of the group onto which the information is to be condensed.

We obtain

$$R_{4,6}(i, j) = P(E(5, 7, 8, 10) \cap G_6(j)|G_4(i))$$
$$= \sum_k \sum_l P(E(5)|G_5(k))P(E(7)|G_7(l))R_7(l)$$
$$\times P(G_6(j)|G_4(i) \cap G_5(k))P(G_7(l)|G_4(i)$$
$$\cap G_5(k))P(G_5(k)), \qquad (6.5.2)$$

which again can be evaluated immediately. The matrix $R_{4,6}$ whose (i, j) element is $R_{4,6}(i, j)$ is given by

$$R_{4,6} = \begin{bmatrix} \tfrac{1}{4}p^2q^2 & \tfrac{1}{4}p^2q^2 & \phi \\ \tfrac{1}{8}pq(p^2+pq) & \tfrac{1}{8}pq(p^2+2pq) & \phi \\ \phi & \phi & \phi \end{bmatrix}.$$

We can now use 3 and 6 jointly. Then

$$R_{3,6}(i, j) = P(E(1, 2, 4, 5, 7, 8, 10) \cap G_3(i) \cap G_6(j))$$
$$= \sum_k P(E(4) \,|\, G_4(k)) R_{4,6}(k, j) P(G_3(i) \cap G_4(k)),$$

where

$$P(G_3(i) \cap G_4(k)) = \sum_j \sum_l P(E(1) \,|\, G_1(j)) P(E(2) \,|\, G_2(l)$$
$$\times P(G_3(i) \cap G_4(k) \,|\, G_1(j)$$
$$\cap G_2(l)) P(G_1(j)) P(G_2(l)).$$

Finally, we obtain

$$R_9(i) = P(E(1, 2, 3, 4, 5, 6, 7, 8, 10) \cap G_9(i))$$
$$= \sum_j \sum_k P(E(3) \,|\, G_3(j)) P(E(6) \,|\, G_6(k)) R_{3,6}(j, k)$$
$$\times P(G_9(i) \,|\, G_3(j) \cap G_6(k)),$$

so

$$\boldsymbol{R}_9 = \frac{p^4 q^2}{32} \begin{bmatrix} 21p^2 + 36pq + 14q^2 \\ 10p^2 + 27pq + 13q^2 \\ p^2 + 5pq + 3q^2 \end{bmatrix}.$$

From this latter expression we can of course find $P(G_9(i) \,|\, E(1, 2, 3, 4, 5, 6, 7, 8, 10))$.

6.5(iii) *Further developments*

It should be clear how the methodology works, although the exact form of the R-functions may not be apparent. Certain of the manipulations carried out involve simply the multiplication of matrices, and in this respect there are similarities to the method of Li and Sacks (1954). The algorithm illustrated above has been implemented in a variety of forms, some yielding algebraic formulae, others requiring that the parameter values be specified (here p and q), and numerical values retained (Thompson (1976e, 1977)). More complicated genetics models, such as linked loci, quantitative traits and multifactorial inheritance can be handled in essentially the same way (Hasstedt and Cartwright, 1979).

In addition to computing overall probabilities, or risks for a single individual, the same methodology provides conditional probabilities jointly for a set of riskees. Similarly it can be used to find the likelihoods of genotypes of ancestors, given information on a current population, and hence the ancestral origin of certain alleles of interest (Thompson, Cannings and Skolnick, 1978). By computing the likelihood that certain original genes were of allelic types no longer present in the population we can thus obtain also extinction probabilities of sets of original founder genes over a specified pedigree. Finally, by labelling relevant founder genes as distinct alleles, a development of the methodology may be used to find the generalized gene-identity-coefficients of Section 6.4 for a set of individuals in an arbitrarily complex relationship (Thompson, 1980*a*).

6.6 Simulation

We have attempted above to demonstrate the workings of a number of algorithms for the calculation of genealogical measures. Theoretically these are capable of handling problems on arbitrarily complex pedigrees, and with arbitrarily complex models. In practice, of course, the amount of computer space and/or time needed may be very large, even when the algorithms have been optimized (something we have not attempted to discuss). In some situations we may turn instead to the use of computer simulation. In this one does not attempt to evaluate specific probabilities, for example the inbreeding coefficient, but rather to estimate them. The computer is used to construct possible realizations of the phenomenon under study, in this case gene-flow in some population or pedigree, and these realizations are treated as a sample, which we may use to estimate required values.

Simulation, or Monte Carlo simulation as it is often called, dates back some one hundred years (see discussion in Hammersley and Handscomb (1964)), but its use in genetics is more recent (see for example, Fraser (1957), Bodmer (1960), Latter (1965) and Crosby (1963, 1973)). Simulation programs vary in the degree of detail which they attempt to incorporate into the underlying model. Broadly we shall distinguish only two categories; the first attempting to model the specific biological interactions of individuals, their matings, births and deaths together with their genetic structure, phenotypes and the effects of the latter on the former; the second considers only the flow of genetic material through an already specified genealogy. In order to distinguish these two categories we will refer to them as population simulations, and genealogical simulations.

It is not appropriate here to discuss in any depth the population simulations. However it is worth making some comment on the level of sophistication of these, and their possible applications, focusing in particular on those modelling human populations. The earliest attempt at such a simulation seems to have been that of Barrai and Barbieri (1964) and Cavalli-Sforza and Zei (1967). Simulations of a similar nature were produced by MacCluer (1967), Skolnick and Cannings (1973), and by Hammel and Hutchinson (1973), though the focus of the last mentioned was sociological rather than genetic. The flow charts of the program of Skolnick and Cannings are shown in Figures 6.6(a) and 6.6(b). Output from these programs will vary according to the focus of the investigators, and the structure of the program. The program of Skolnick and Cannings will produce, if required, a complete genealogy of all the individuals created during the simulation, and this may then be used to calculate any of the measures of interest.

We turn now to what we have called genealogical simulations, i.e. those which take a given genealogy, and then simulate the flow of genetic material through it. Edwards (1969) developed such a program in which a single autosomal locus was considered, and Skolnick and Cannings (1973) produced a version which ran many loci simultaneously, essentially, since there was no linkage, as a means of economizing in the production of the many replicates needed. Recently Bishop and Cannings (1979) have written a program capable of handling complex genetic systems, the limitations being imposed by computer space and time, rather than by the nature of the algorithm. The algorithm uses the bit structure of the computer word to code for each allele, and the random segregation is achieved by using the bits of a randomly generated number. The program will accept any genealogy, and the user can construct his desired genetic system, including multiple loci and sex-linked loci, provided that it is diploid. There is no facility for polyploidy.

We illustrate the use of such a program in the present context by reference to the pedigree shown in Figure 6.6(c). Suppose we wish to investigate the genetic relationship of 9 and 10 for a single autosomal locus, who are double-first-cousins. We have already

Figure 6.6(*a*). Flow diagram of individual's life. (From Skolnick and Cannings (1973).)

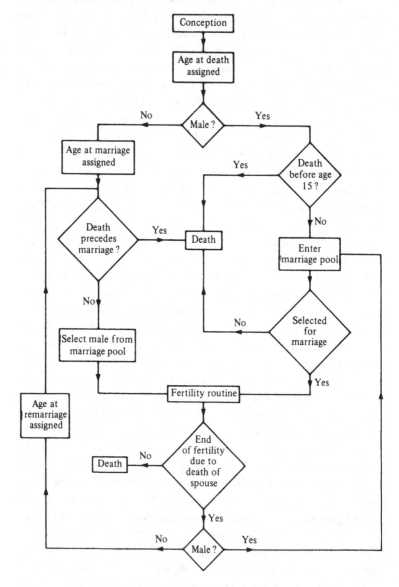

Figure 6.6(*b*). Flow diagram for fertility routine. (From Skolnick and Cannings (1973).)

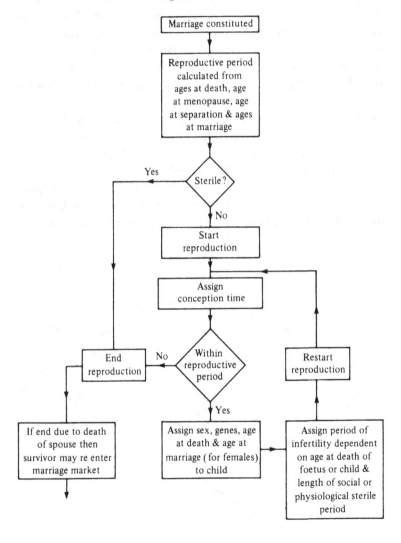

seen earlier (Section 2.7) that, relative to the founders 1, 2, 3 and 4, i.e. assuming they have no genes identical by descent, the only identity states for 9 and 10 are $(1, 2, 1, 2)$, $(1, 2, 1, 3)$ and $(1, 2, 3, 4)$, and that the probabilities are $\frac{1}{16}$, $\frac{6}{16}$ and $\frac{9}{16}$, respectively. We can infer the coefficient of kinship to be $\frac{1}{8}$, and the coefficient of inbreeding to be 0. Supposing that we had no such algorithmic methods available we could proceed as follows, using the simulation procedure: (i) assign alleles 1 to 8 to the individuals 1, 2, 3 and 4; (ii) similate the gene-flow through the genealogy to obtain an array of alleles for 9 and 10; (iii) replace this by its orbit. We repeat this process a large number of times, n say, obtaining frequencies n_1, n_2 and $n_3(n_1 + n_2 + n_3 = n)$ for the three possible identity states. The estimation of the probabilities of those identity states, and of the coefficients of kinship and inbreeding, are then straightforward (the estimates are n_1/n, n_2/n, n_3/n, $(2n_1 + n_2/2n)$ and 0). Standard errors for these estimates are straightforward to obtain, and any desired degree of precision can be achieved, though, as pointed out by Edwards, care

Figure 6.6(c).

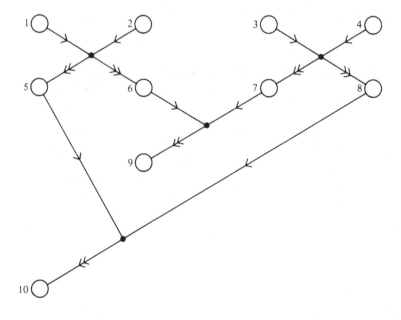

must be taken to allow for the limited length of the sequence of pseudo-random numbers available on one's computer.

If one wishes to study a regular system of mating then it is a simple matter to identify the individual at the bottom of one's genealogy, with a new set to be placed at the top of the next round of simulation.

REFERENCES

Atkins, J. R. (1974a). Grafik: a multipurpose kinship metalanguage. In *Proceedings of the Math. Soc. Sci. Board Conference on Genealogical Mathematics*, P. Ballonoff (ed.). The Hague: Mouton and Co.

Atkins, J. R. (1974b). Consanguineal distance measures: a mathematical analysis. In *Mathematical Models of Social and Cognitive Structures. Contributions to the Mathematical Development of Anthropology*, P. Ballonoff (ed.). Urbana, Illinois: University of Illinois Press.

Barrai, I. and Barbieri, D. (1964). Drift in una populazione simulata. *Atti. Ass. Genet.*, **9**, 233–45.

Bishop, D. T. and Cannings, C. (1979). Sim: A simulation program for studying gene flow through a pedigree. Technical Report No. 11, Department of Medical Biophysics and Computing, University of Utah.

Bodmer, W. F. (1960). The genetics of homostyly in populations of Primula Vulgaris. *Phil. Trans. Roy. Soc. (Lond.)*, **242B**, 517–19.

Bodmer, W. F. and Cavalli-Sforza, L. L. (1968). A migration matrix model for the study of random genetic drift. *Genetics*, **59**, 565–92.

Campbell, M. A. and Elston, R. C. (1971). Relatives of probands; models for preliminary genetic analysis. *Ann. Hum. Genet. (Lond.)*, **35**, 225–36.

Cannings, C. (1973). The equivalence of some overlapping and non-overlapping models for the study of genetic drift. *J. Appl. Prob.*, **10**, 432–6.

Cannings, C. (1974a). The latent roots of certain Markov chains arising in genetics; a new approach. I. Haploid models. *Adv. Appl. Prob.*, **6**, 260–90.

Cannings, C. (1974b). Genetic drift with polygamy and arbitrary offspring distribution. *J. Appl. Prob.*, **11**, 633–41.

Cannings, C. (1975). The latent roots of certain Markov chains arising in genetics: a new approach. II. Further haploid models. *Adv. Appl. Prob.*, **7**, 264–82.

Cannings, C. and Skolnick, M. H. (1975). Genetic drift in exogamous marriage systems. *Theor. Pop. Biol.*, **7**, 39–54.

Cannings, C., Thompson, E. A. and Skolnick, M. H. (1976). The recursive derivation of likelihood on complex pedigrees. *Adv. Appl. Prob.*, **8**, 622–5.

Cannings, C., Thompson, E. A. and Skolnick, M. H. (1978). Probability functions on complex pedigrees. *Adv. Appl. Prob.* **10**, 26–61.

Cannings, C., Thompson, E. A. and Skolnick, M. H. (1979). Extension of pedigree analysis to include assortative mating and linear models. In *Genetic Analyses of Common Diseases: Applications to Predictive Factors in Coronary Disease*, C. F. Sing & M. H. Skolnick (eds.). New York: Alan R. Liss, Inc.

Cannings, C., Thompson, E. A. and Skolnick, M. H. (1980). Pedigree analysis of complex models. In *Current Developments in Anthropological Genetics*, J. Mielke & M. Crawford, (eds.). New York: Plenum Press.

Carnap, R. (1958). Introduction to Symbolic Logic and its Applications. New York: Dover.

Cavalli-Sforza, L. L. (1966). Population structure and human evolution. *Proc. Roy. Soc. (Lond.)*, **164B**, 362–79.

Cavalli-Sforza, L. L. and Edwards, A. W. F. (1967). Phylogenetic analysis: models and estimation procedures. *Amer. J. Hum. Genet.*, **19**, 233–57.

Cavalli-Sforza, L. L. and Zei, G. (1966). Experiments with an artificial population. *Proceedings of the Third International Congress of Human Genetics*. Chicago: University of Chicago Press.

Chakraborty, R., Shaw, M., and Schull, W. J. (1974). Exclusion of paternity: the current state of the art. *Amer. J. Hum. Genet.*, **26**, 477–88.

Chapman, A. M. and Jacquard, A. (1971). Un isolat d'Amerique centrale: Les Indiens Jicaques du Honduras. In *Genetique et Population: Homage a Jean Sutter*. Paris: Presses Univ. de France.

Chia, A. B. (1968). Random mating in a population of cyclic size. *J. Apppl. Prob.* **5**, 21–30.

Chia, A. B. and Watterson, G. A. (1969). Demographic effects on the rate of genetic evolution. I. Constant size populations with two genotypes. *J. Appl. Prob.*, **6**, 231–49.

Cockerham, C. C. (1956). Effects of linkage on the covariances between relatives. *Genetics*, **41**, 1141–8.

Cockerham, C. C. (1967). Group inbreeding and coancestry. *Genetics*, **56**, 89–104.

Cockerham, C. C. (1969). Variance of gene frequencies. *Evolution*, **23**, 72–84.

Cockerham, C. C. (1971). Higher order probability functions of identity of alleles by descent. *Genetics*, **69**, 235–46.

Cockerham, C. C. (1973). Analyses of gene frequencies. *Genetics*, **74**, 679–700.

Cotterman, C. W. (1940). A calculus for statistico genetics. Ph.D. Thesis Ohio State University. Published in *Genetics and Social Structure*, P. Ballonoff (ed.). (1975). Benchmark papers in Genetics, Dowden, Hutchinson and Ross.

Crosby, J. L. (1963). Evolution by computer. *New Scientist*, **327**, 415–17.

Crosby, J. L. (1973). Computer Simulation in Genetics. New York: Wiley.

Crow, J. F. and Kimura, M. (970). An Introduction to Population Genetics Theory. New York: Harper & Row.

Denniston, C. (1975). Probability and genetic relationship: two loci. *Ann. Hum. Genet. (Lond.)*, **39**, 89–104.

Edwards, A. W. F. (1967). Automatic construction of genealogies from phenotypic information (Autokin). *Bull. Eur. Soc. Hum. Genet.*, **1**, 42–43.

Edwards, A. W. F. (1969). Discussion to a paper by A. P. Mange. In *Computer Applications in Genetics*, N. E. Morton (ed.). Honolulu: University Press of Hawaii.

Edwards, A. W. F. (1970). Estimation of the branch points of a branching-diffusion process. *J. Roy. Statist. Soc.* **32B**, 154–74.

Edwards, A. W. F. (1979). Parsing a genealogy. *Adv. Appl. Prob.*, **11**, 2–3.

Edwards, A. W. F. and Cavalli-Sforza, L. L. (1963). The reconstruction of evolution. *Heredity*, **18**, 553.

Elston, R. C. and Lange, K. (1976). The genotypic distribution of relatives of homozygotes when consanguinity is present. *Ann. Hum. Genet. (Lond.).*, **39**, 493–96.

Elston, R. C. and Stewart, J. (1971). A general model for the genetic analysis of pedigree data. *Human Heredity*, **21**, 523–42.

Ewens, W. J. (1973). Testing for increased mutation rate for neutral alleles. *Theor. Pop. Biol.*, **4**, 251–8.

Falconer, D. S. (1960). Introduction to Quantitative Genetics. New York: Ronald Press Co.

Feller, W. (1951). Diffusion processes in genetics. In *Proceedings Second Berkeley Symposium*, pp. 227–46, J. Neyman (ed.). Berkeley, California: University of California Press.

Feller, W. (1968). A First Course in Probability Theory and its Applications. Vol. 1 (3rd. edn). New York: Wiley.

Felsenstein, J. (1968). Statistical Inference and the Estimation of Phylogenies. Ph.D. Thesis, University of Chicago.

Felsenstein, J. (1971*a*). The rate of loss of multiple alleles in finite haploid populations. *Theor. Pop. Biol.*, **2**, 391–403.

Felsenstein, J. (1971*b*). Inbreeding and variance effective numbers in populations with overlapping generations. *Genetics*, **68**, 581–97.

Felsenstein, J. (1973). Maximum likelihood estimation of trees from continuous characters. *Amer. J. Hum. Genet.*, **25**, 471–92.

Fisher, R. A. (1930). The Genetical Theory of Natural Selection. Oxford: Clarendon Press.

Fisher, R. A. (1949). The Theory of Inbreeding. Edinburgh: Oliver and Boyd.

Fitch, W. M. and Margoliash, E. (1967). Construction of evolutionary trees. *Science*, **155**, 279–84.

Fox, R. (1967). Kinship and Marriage. London: Pelican.

Fraser, A. S. (1957). Simulation of genetic systems by automatic digital computer. I. Introduction. *Aust. J. Biol. Sci.*, **10**, 484–91.

Fraser Roberts J. A. and Pembry M. E. (1978). An Introduction to Medical Genetics. (7th. ed). Oxford: Oxford University Press.

Gallais, A. (1974). Covariances between arbitrary relatives with linkage and epistasis, in the case of linkage disequilibrium. *Biometrics* **30**, 429–446.

Gillois, M. (1964). La Relation d'Identite en Genetique. Thesis, University of Paris.

Gillois, M. (1965). Relation d'identite en genetique. *Ann. Inst. Henri Poicare*, **2B**, 1–94.

Goody, J. (ed.). (1971). Kinship. London:Penguin.

Gray Eaton, G. (1976). The social order of Japanese macaques. *Sci. Amer.*, **235**, 97–106.

Griffiths, R. C. (1980). Lines of descent in the diffusion approximation of neutral Wright–Fisher models. *Theor. Pop. Biol.*, **17**, 37–50.

Haldane, J. B. S. (1949). The association of characters as a result of inbreeding and linkage. *Ann. Eugen.*, **15**, 15–23.

Haldane, J. B. S. and Jayakar, S. D. (1962). An enumeration of some human relationships. *J. Genet.*, **58**, 81–107.

Hammel, E. A. and Hutchinson, D. (1973). Two tests of computer microsimulation: The effect of an incest taboo on population viability, and the effect of age differences between spouses on the skewing of consanguineal relationships between them. In *Computer Simulation in Human Population Studies*, B. Dyke & J. W. MacCluer (eds.). New York: Academic Press.

Hammersley, J. M. and Handscomb, D. C. (1964). Monte Carlo Methods. London: Methuen.

Harris, D. L. (1964). Genotypic covariances between inbred relatives. *Genetics*, **50**, 1319–48.

Hasstedt, S. and Cartwright, P. (1979). PAP – Pedigree Analysis Package. Technical Report No. 13, Department of Medical Biophysics and Computing, University of Utah.

Heuch, I. and Li F. H. F. (1972). PEDIG – A computer program for calculation of genotype probabilities using phenotype information. Clinical Genetics, **3**, 501–4.

Hilden, J. (1970). GENEX – An algebraic approach to pedigree probability calculus. *Clinical Genetics*, **1**, 319–48.

Jacquard, A. (1971). Effect of exclusion of sib mating on genetic drift. *Theor. Pop. Biol.*, **2**, 91–9.

Jacquard, A. (1972). Genetic information given by a relative. *Biometrics*, **28**, 1101–14.

Jacquard, A. (1974). The Genetic Structure of Populations. New York: Springer-Verlag.

Jacquard, A. (1975). Inbreeding: one word; several meanings. *Theor. Pop. Biol.*, **7**, 338–63.

Kahane, J. and Marr, R. (1972). On a class of stochastic pairing processes and the Mycielski-Ulam notions of genealogical distance. *J. Combinatorial Theory*, **13A**, 383–400.

Karlin, S. (1966). A First Course in Stochastic Processes. New York: Academic Press.

Karlin, S. (1968*a*). Rates of approach to homozygosity for finite stochastic models with variable population size. *Amer. Nat.*, **102**, 443–55.

Karlin, S. (1968*b*). Equilibrium behaviour of population genetic models with non-random mating. I. Preliminaries and special mating systems. *J. Appl. Prob.*, **5**, 231–13.

Karlin, S. (1968*c*). Equilibrium behaviour of population genetic models with non-random mating. II. Pedigrees, homozygosity and stochastic models. *J. Appl. Prob.*, **5**, 487–566.

Karlin, S. and McGregor, J. (1964). Direct product branching processes and related Markov chains. *Proc. Nat. Acad. Sci. (USA)*, **51**, 598–602.

Karlin, S. and McGregor, J. (1965). Direct product branching processes and related induced Markov chains. I. Calculations of rates of approach to homozygosity. In *Bernoulli, Bayes, Laplace Anniversary Volume*. Berlin: Springer-Verlag.

Karlin, S. and Taylor H. M. (1975). A First Course in Stochastic Processes. (2nd. edn). London: Academic Press.

Kempthorne, O. (1957). An Introduction to Genetic Statistics. New York: Wiley.

Kempthorne, O. (1968). The concept of identity of genes by descent. In *Proceedings Fifth Berkeley Symposium*. J. Neyman (ed.). Berkeley, California: University of California Press.

Kendall, D. G. (1971). The algebra of genealogy. *Math. Spectrum*, **4**, 7–8.

Kimura, M. (1955). Random genetic drift in a multiallelic locus. *Evolution*, **9**, 419–35.

Kimura, M. and Crow, J. F. (1963). On the maximum avoidance of inbreeding. *Genetical Research*, **4**, 399–415.

Kimura, M. and Weiss, G. H. (1964). The stepping stone model of population structure and the decrease of genetic correlation with distance. *Genetics*, **49**, 561–76.

Kruskal, J. B. and Wish, M. (1978). Multidimensional Scaling. Beverly Hills: Sage Publications.

Lange, K. (1974). Relative-to-relative transition probabilities for two linked genes. *Theor. Pop. Biol.*, **6**, 92–107.

Latter, B. D. H. (1965). The response to artificial selection due to autosomal genes of large effect. I. The effects of linkage on limits to selection in finite populations. *Aust. J. Biol. Sci.*, **18**, 1009–23.

Levi-Strauss, C. (1968). Structural Anthropology. London: Penguin.

Lewontin, R. C. and Krakauer, J. (1973). Distribution of gene frequency as a test of the theory of selective neutrality of polymorphisms. *Genetics*, **74**, 175–95.

Li, C. C. and Sacks, L. (1954). The derivation of joint distribution and correlation between relatives by the use of stochastic matrices. *Biometrics*, **10**, 347–60.

Littler, R. A. (1975). Loss of variability at one locus in a finite population. *Math. Biosci.*, **25**, 151–63.

Lloyd, S. P. (1973). An index of genealogical relatedness. *Adv. Appl. Prob.*, **5**, 417–38.

Lloyd, S. P. and Mallows, C. L. (1973). An index of genealogical relatedness derived from a genetic model. Ann. Prob., **1**, 758–71.

MacCluer, J. W. (1967). Monte Carlo methods in human population genetics: A computer model incorporating age-specific birth and death rates. *Amer. J. Hum. Genet.*, **19**, 303–12.

Mair, L. (1971). Marriage. London:Penguin.

Malecot, G. (1948). Les mathematiques de l'heredite. Paris:Masson et Cie.

Mallows, C. L. (1974). Another definition of genealogical distance. *J. Appl. Prob.*, **11**, 179–83.

Maruyama, T. (1969). On the fixation probability of mutant genes in a subdivided population. *Genetical Research*, **15**, 221–5.

Maruyama, T. (1970a). Analysis of population structure. I. One-dimensional stepping-stone models of finite length. *Ann. Hum. Genet. (Lond.)*, **34**, 201–19.

Maruyama, T. (1970b). Stepping-stone models of finite length. *Adv. Appl. Prob.*, **2**, 229–58.

Maruyama, T. (1970c). The rate of decrease of heterozygosity in a population occupying a circular or linear habitat. *Genetics*, **67**, 437–54.

Moran, P. A. P. (1958). Random processes in genetics. *Proc. Camb. Phil. Soc.*, **54**, 60–71.

Moran, P. A. P. (1959). The theory of some genetical effects of population subdivision. *Aust. J. Biol. Sci.*, **12**, 109–18.

Moran, P. A. P. (1962). The Statistical Processes of Evolutionary Theory. Oxford: Oxford University Press.

Moran, P. A. P. and Watterson, G. A. (1959). The genetic effects of family structure in natural populations. *Aust. J. Biol. Sci.*, **12**, 1–15.

Morton, N. E., Yasuda, N., Miki, C. and Yee, S. (1968). Bioassay of population structure and isolation by distance. *Amer. J. Hum. Genet.*, **20**, 411–19.

Morton, N. E., Yee, S., Harris, D. E. and Lew, R. (1971). Bioassay of kinship. *Theor. Pop. Biol.*, **2**, 507–24.

Mycielski, J. and Ulam, S. M. (1969). On the pairing process and the notion of genealogical distance. *J. Combinatorial Theory*, **6**, 227–34.

Nadot, R. and Vaysseix, G. (1973). Apparentement et identite: Algorithme du calcul des coefficients d'identite. *Biometrics*, **29**, 347–59.

Neel, J. V. (1970). Lessons from a 'primitive' people. *Science*, **170**, 815–22.

Neel, J. V. and Ward, R. H. (1970). Village and tribal genetic distances among American Indians, and the possible implications for human evolution. *Proc. Nat. Acad. Sci. (USA)*, **65**, 323–30.

Nei, M. (1972). A new measure of genetic distance. In *Genetic Distance*, J. F. Crow (ed.). New York: Plenum Press.

Nei, M. (1973). The theory and estimation of genetic distance. In *Genetic Structure of Populations*, N. E. Morton (ed.). Honolulu: University Press of Hawaii.

Nei, M. and Chakravarti, A. (1977). Drift variances of FST and GST statistics obtained from a finite number of isolated populations. *Theor. Pop. Biol.*, **11**, 307–25.

Nei, M., Chakravarti, A. and Tateno, Y. (1977). Mean and variance of FST in a finite number of incompletely isolated populations. *Theor. Pop. Biol.*, **11**, 291–306.

Nei, M. and Roychoudhury, A. K. (1972). Gene differences between Caucasian, Negro and Japanese populations. *Science*, **177**, 434–436.

Piva, M. and Holgate, P. (1977). The eigenvalues of a finite population model. *Ann. Hum. Genet. (London.)*, **41**, 103–6.

Roberts, D. F. (1971). The demography of Tristan da Cunha. *Pop. Stud.* **25**, 465–79.

Robertson, A. (1952). The effect of inbreeding on the variation due to recessive genes. *Genetics*, **37**, 189–207.

Robertson, A. (1964). The effect of non-random mating within inbred lines on the rate of inbreeding. *Genetical Research*, **5**, 164–67.

Schnell, F. W. (1963). The covariances between relatives in the presence of linkage. In *Statistical Genetics and Plant Breeding*, p. 468–83, W. D. Hanson & H. F. Robinson (eds.). New York: Humphrey.

Seal, H. (1964). Multivariate statistical analysis for biologists. London: Methuen.

Skolnick, M. H. and Cannings, C. (1973). Simulation of small human populations. In *Computer Simulation in Human Population Studies*, B. Dyke and J. W. MacCluer (eds.). New York: Academic Press.

Smith, C. A. B. (1976). The use of matrices in calculating Mendelian probabilities. *Ann. Hum. Genet. (Lond.)*, **41**, 117–21.

Sokal, R. R. and Sneath, P. H. A. (1963). Principles of Numerical Taxonomy. San Francisco: Freeman.

Stevens, A. (1975). An elementary computer algorithm for calculation of the coefficient of inbreeding. *Inf. Proc. Lett.*, **3**, 153–63.

Thompson, E. A. (1974*a*). Gene identities and multiple relationships. *Biometrics*, **30**, 667–80.

Thompson, E. A. (1974*b*). Mathematical Analysis of Human Evolution and Population Structure. Ph.D. Thesis, University of Cambridge.

Thompson, E. A. (1975*a*). Estimation of pairwise relationships. *Ann. Hum. Genet. (Lond.)*, **39**, 173–88.

Thompson, E. A. (1975*b*). Human Evolutionary Trees. Cambridge: Cambridge University Press.

Thompson, E. A. (1976*a*). Population correlation and population kinship. *Theor. Pop. Biol.*, **10**, 205–26.

Thompson, E. A. (1976*b*). A restriction on the space of genetic relationships. *Ann. Hum. Genet. (Lond.)*, **40**, 201–4.

Thompson, E. A. (1976*c*). A paradox of genealogical inference. *Adv. Appl. Prob.*, **8**, 648–50.

Thompson, E. A. (1976*d*). Inference of genealogical structure. *Soc. Sci. Inform.*, **15**, 477–526.

Thompson, E. A. (1976*e*). Peeling programs for zero-loop pedigrees. Technical Report No. 5, Department of Medical Biophysics and Computing, University of Utah.

Thompson, E. A. (1977). Peeling programs for pedigrees of arbitrary complexity. Technical Report No. 6, Department of Medical Biophysics and Computing, University of Utah.

Thompson, E. A. (1978). Impossible gene identity states. *Adv. Appl. Prob.*, **10**, 19–22.

Thompson, E. A. (1979*a*). Genealogical structure and correlations in gene extinction. *Theor. Pop. Biol.*, **16**, 191–222.

Thompson, E. A. (1979*b*). Ancestral inference. III. The ancestral structure of the population of Tristan da Cunha. *Ann. Hum. Genet. (Lond.)*, **43**, 167–76.

Thompson, E. A. (1980*a*). Recursive routines for pedigree analysis. Technical Report No. 17, Department of Medical Biophysics and Computing, University of Utah.

Thompson, E. A. (1980*b*). The gene identity states of a descendant. *Theor. Pop. Biol.*, **18**, 76–93.

Thompson, E. A. (1980*c*). Genetic etiology and clusters in a pedigree. *Heredity*, **45**, 323–34.

Thompson, E. A., Cannings, C. and Skolnick, M. H. (1978). Ancestral inference. I. The problem and the method. *Ann. Hum. Genet. (Lond.)*, **42**, 95–108.

Thompson, E. A. and Neel, J. V. (1978). The probability of founder effect in a tribal population. *Proc. Nat. Acad. Sci. (USA)*, **75**, 1442–5.

Trustrum, G. B. (1961). The correlations between relatives in a random mating diploid population. *Proc. Camb. Phil. Soc.*, **57**, 315–20.

van Aarde, I. M. R. (1975). The covariance of relatives derived from a random mating population. *Theor. Pop. Biol.*, **8**, 166–83.

Wahlund, S. (1928). Zuzammensetzung von Population und Korrelationserscheinungen vom Standpunkt der Verebungslehre aus betrachtet. Hereditas, **11**, 65–106.

Weiss, G. H. and Kimura, M. (1965). A mathematical analysis of the stepping stone model of genetic correlation. *J. Appl. Prob.*, **2**, 129–49.

Wilson, E. O. (1975). Sociobiology. Cambridge, Mass., U.S.A.: Belnap Press.

Wright, S. (1921). Systems of mating. *Genetics*, **6**, 111–78.

Wright, S. (1922). Coefficients of inbreeding and relationship. *Amer. Nat.*, **56**, 330–38.

Wright, S. (1931). Evolution in Mendelian populations. *Genetics*, **16**, 97–159.

Wright, S. (1933). Inbreeding and homozygosis. *Proc. Nat. Acad. Sci. (USA)*, **19**, 411–20.

Wright, S. (1940). Breeding structure of populations in relation to speciation. *Amer. Nat.*, **74**, 232–48.

Wright, S. (1943). Isolation by distance. *Genetics*, **28**, 114–38.

Wright, S. (1946). Isolation by distance under diverse systems of mating. *Genetics*, **31**, 39–59.

Wright, S. (1951). The genetical structure of populations. *Ann. Eugen.*, **15**, 323–54.

Wright, S. (1965). The interpretation of population structure by F-statistics with special regard to systems of mating. *Evolution*, **19**, 395–420.

AUTHOR INDEX

The numbers refer to sections.

SUBJECT INDEX

The numbers refer to sections.